MUNICIPAL SOLID WASTE ENERGY CONVERSION IN DEVELOPING COUNTRIES

MUNICIPAL SOLID WASTE ENERGY CONVERSION IN DEVELOPING COUNTRIES

Technologies, Best Practices, Challenges and Policy

Edited by

SUANI TEIXEIRA COELHO (Editor-in-Chief)

ALESSANDRO SANCHES PEREIRA (co-Editor-in-Chief)

DANIEL HUGO BOUILLE (Regional Editor)

SHYAMALA K. MANI (Regional Editor)

MARINA YESICA RECALDE (Regional co-Editor)

ATILIO ARMANDO SAVINO (Regional co-Editor)

WILLIAM H.L. STAFFORD (Regional Editor)

ELSEVIER

Elsevier
Radarweg 29, PO Box 211, 1000 AE Amsterdam, Netherlands
The Boulevard, Langford Lane, Kidlington, Oxford OX5 1GB, United Kingdom
50 Hampshire Street, 5th Floor, Cambridge, MA 02139, United States

Notices
Knowledge and best practice in this field are constantly changing. As new research and experience broaden our understanding, changes in research methods, professional practices, or medical treatment may become necessary.

Practitioners and researchers must always rely on their own experience and knowledge in evaluating and using any information, methods, compounds, or experiments described herein. In using such information or methods they should be mindful of their own safety and the safety of others, including parties for whom they have a professional responsibility.

To the fullest extent of the law, neither the Publisher nor the authors, contributors, or editors, assume any liability for any injury and/or damage to persons or property as a matter of products liability, negligence or otherwise, or from any use or operation of any methods, products, instructions, or ideas contained in the material herein.

Library of Congress Cataloging-in-Publication Data
A catalog record for this book is available from the Library of Congress

British Library Cataloguing-in-Publication Data
A catalogue record for this book is available from the British Library

ISBN: 978-0-12-813419-1

For information on all Elsevier publications
visit our website at https://www.elsevier.com/books-and-journals

Publisher: Joe Hayton
Acquisition Editor: Raquel Zanol
Editorial Project Manager: Joanna Collett
Production Project Manager: Mohana Natarajan
Cover Designer: Matthew Limbert

Typeset by SPi Global, India

Working together
to grow libraries in
developing countries

www.elsevier.com • www.bookaid.org

Contents

Contributors

Babu J. Alappat
Indian Institute of Technology, New Delhi, India

Daniel Hugo Bouille
Fundación Bariloche, Río Negro, Argentina

Alejandro Cittadino
Departamento de Ecología, Genética y Evolución—Facultad de Ciencias Exactas y Naturales, Universidad de Buenos Aires, Buenos Aires, Argentina

Suani Teixeira Coelho
Research Group on Bioenergy, Institute of Energy and Environment, University of São Paulo, São Paulo, Brazil

Rocio Diaz-Chavez
SEI Africa, Stockholm, Sweden

Javier Farago Escobar
Research Group on Bioenergy, Institute of Energy and Environment, University of São Paulo, São Paulo, Brazil

Vanessa Pecora Garcilasso
Research Group on Bioenergy, Institute of Energy and Environment, University of São Paulo, São Paulo, Brazil

Luciano Infiesta
Carbogas Industries, São Paulo, Brazil

Caio Luca Joppert
Research Group on Bioenergy, Institute of Energy and Environment, University of São Paulo, São Paulo, Brazil

Dinesh Kumar
Indian Institute of Technology, New Delhi, India

Shyamala K. Mani
National Institute of Urban Affairs (NIUA), India Habitat Centre, New Delhi, India

Max Mapako
Council for Scientific and Industrial Research, Pretoria, South Africa

Madzore Mapako
Council for Scientific and Industrial Research, Pretoria, South Africa

Juan Daniel Martínez
Grupo de Investigaciones Ambientales (GIA), Universidad Pontificia Bolivariana (UPB), Medellín, Colombia

Ben Muok
Centre for Research, Innovation and Technology, Jaramogi Odinga University, Nairobi, Kenya, East Africa

Suzan Oelofse
Council for Scientific and Industrial Research, Pretoria, South Africa

Fernando C. de Oliveira
Research Group on Bioenergy, Institute of Energy and Environment, University of São Paulo, São Paulo, Brazil

Walter Ospina
Consejo de Investigación y Tecnologí as de Valorización Energética de Residuos de Colombia (WTERT—Colombia), Colombia, SC, United States of America

Suneel Pandey
The Energy and Resources Institute (TERI), New Delhi, India

Dinesh Chandra Pant
The Energy and Resources Institute (TERI), New Delhi, India

Agamuthu Pariatamby
University of Malaya, Kuala Lumpur, Malaysia

Alessandro Sanches Pereira
Instituto 17, São Paulo, Brazil

Osvaldo Soliano Pereira
Universidade Federal da Bahia—UFBA (Federal University of Bahia), Salvador, Brazil

Marina Yesica Recalde
Fundación Bariloche, Buenos Aires, Argentina

Enrique Posada Restrepo
Área de Innovación y Desarrollo Hatch, Portland, OR, United States of America

Albert Rugumayo
Faculty of Engineering, Ndejje University of Uganda, Kampala, East Africa

Laura Salgado
Departamento de Ingeniería Mecánica, Escuela Politécnica Nacional, Ladrón de Guevara, Quito, Ecuador

Bini Samal
Forest Research Institute, Dehradun, India

Estela Santalla
Departamento Ingeniería Química, Facultad de Ingeniería/UNICEN, Buenos Aires, Argentina

Marilin Mariano dos Santos
Research Group on Bioenergy, Institute of Energy and Environment, University of São Paulo, São Paulo, Brazil

Atilio Armando Savino
International Solid Waste Association (ISWA), Buenos Aires, Argentina

Roshni Mary Sebastian
Indian Institute of Technology, New Delhi, India

Pratibha Sharma
Global Alliance for Incinerator Alternatives, Pune, India

José R. Simões-Moreira
Polytechnic School, University of São Paulo, São Paulo, Brazil

Fábio Rubens Soares
Research Group on Bioenergy, Institute of Energy and Environment, University of São Paulo, São Paulo, Brazil

Gustavo Solórzano
Asociación Mexicana de Ingeniería, Ciencia y Gestión Ambiental, A.C. (AMICA), Mexico City, Mexico

Rafael Soria
Departamento de Ingeniería Mecánica, Escuela Politécnica Nacional, Ladrón de Guevara, Quito, Ecuador

William H.L. Stafford
Council for Scientific and Industrial Research; Department of Industrial Engineering, University of Stellenbosch, Stellenbosch, South Africa

Luís Gustavo Tudeschini
Research Group on Bioenergy, Institute of Energy and Environment, University of São Paulo, São Paulo, Brazil

About the Editors

Suani Teixeira Coelho (Editor-in-Chief) acquired her M.Sc. and Ph.D. in Energy at Universidade de São Paulo (USP), where she is a professor in the Energy Post-Graduate Program (PPGE). She is also a professor at the joint Ph.D. Program in Bioenergy of USP, Universidade de Campinas (Unicamp), and Universidade Estadual Paulista (Unesp). She coordinates GBIO Bioenergy Research Group of USP's Energy and Environment Institute, conduction research primarily in bio-  mass and biomass energy generation, distributed energy generation, municipal and rural solid waste, cogeneration, biogas, life cycle analysis, external factors, and sugarcane. She is a recurring reviewer of several energy journals, including *Energy Policy* and *Biomass and Bioenergy*. She is also Bioenergy editor for *Renewable and Sustainable Energy Reviews*.

Alessandro Sanches Pereira (co-Editor-in-Chief) has a B.Sc. in Sanitation Technology from State University of Campinas (Unicamp), M.Sc. in Environmental Management and Policy from the University of Lund, Sweden, Ph.D. in Civil Engineering in the area of Sanitation and Environment from Unicamp, and a Postdoctoral Degree in Energy Planning from the University of São Paulo (USP). Alessandro has worked as a researcher at the Royal Institute of Technology (KTH), Sweden, and as a consultant for the United Nations Conference on  Trade and Development (UNCTAD). He currently is the Executive Director of the Instituto 17, a nonprofit organization, founded in 2018, to enable the dissemination of the sustainable development objectives (SDGs) and propose solutions based on circular economy, environmental protection, and local development. He is associate researcher at the Bioenergy Research Group (GBio) and the Research Centre for Gas Innovation (RCGI). Alessandro is also one of the Brazilian researchers who participated in Group III of the Intergovernmental Panel on Climate Change (IPCC) for the preparation of the AR6 report.

Daniel Hugo Bouille (Regional Editor)

- *Economist—National University of Rosario (Argentina)— Postgraduate studies in Energy Economics Institute— University of Cologne—Germany.*
- *Bariloche Foundation—Senior Researcher and Head of the department of environment and development.*
- *UNFCCC—IPCC—Coordinator Lead Author WGIII— 3AR.*
- *UNFCCC—IPCC—Member of the "Task Group on Data and Scenario Support for impact and Climate Analysis" (TGICA).*
- *UNFCCC—IPCC—WG III—Review Editor—5AR—Chapter III*
- *UNDP-Climate Change—National Communication Support Program—Member of the Roster of Experts*
- *Climate Technology Center and Network (CTCN)—Representant of Bariloche Foundation*

Short Bio: Daniel Hugo Bouille

He is an economist, postgraduate in economics and energy and environmental policy, and fluent in Spanish and English.

- Since 1974, he has developed his entire professional career as an expert in economics and energy planning based in Argentina, with an exception in 1978/80 when he completed his postgraduate studies in Germany.
- In-depth knowledge of all aspects related to the countries energy policy and planning: institutional, legislation, energy policy, tariffs, financing, sectoral analysis (oil, natural gas, hydropower, nuclear, biomass, renewable) and the energy efficiency programs formulated in Argentina.
- He has been general coordinator of the Second National Communication to UNFCCC of Argentina and team leader of the GHG inventory and energy actions in mitigation.
- He has been a speaker at numerous seminars on the energy situation in LA&C Countries.
- Responsible for the elaboration and coordination of energy action plans, energy efficiency roadmaps and long-term strategies for the energy sector in different countries in Latin America.
- Acted as Team Leader in various studies and projects, among others: "Future energy and power matrix definition of Venezuela" or "Energy Planning Guidelines implementation: Bolivia and Honduras."
- Technical Expert in studies on energy plan elaboration: "Study to development the strategy or the energy sector—Peru," "Strategic Long-Term Plan—Argentina," or "Prospective Study of the Energy System. Proposal of an energy plan: information, scenarios, prospective, and models—Peru"

- He has 44 years of experience in the energy sector developing numerous studies and providing technical assistance on energy economics, energy policy planning, energy efficiency, and climate change in Argentina and several LAC countries.
- Extensive experience in Latin America and Caribbean: Bolivia, El Salvador, Colombia, Ecuador, México, Paraguay, Peru, Uruguay, Venezuela, Costa Rica, Guatemala, Nicaragua, Panamá, Honduras, Cuba, Barbados, Jamaica, Trinidad and Tobago.
- Extensive experience in the design, planning, and implementation of public policies related to energy and specifically energy efficiency and sustainable energy.
- Responsible for the development and implementation of policies and incentive schemes to increase energy efficiency in industry, transport, buildings among others.
- Project leader and technical experts in studies and projects with the focus on providing technical assistance for the planning and implementation of energy efficiency action plans: "Implementation of a roadmap for energy efficiency actions and instruments in the industrial sector in México. Diagnosis, identification of measures, definition of strategic lines and instruments to improve energy efficiency,"
- Technical Assistance to LA&C—Coordinator in Colombian Activities on Energy Efficiency—PROURE evaluation and proposal of new actions.
- Sustainable Energy for All—Report on Energy Efficiency Actions and policies in LA&C.
- Proposal for the implementation of an integrated strategy for energy efficiency in Uruguay

Dr. Shyamala K. Mani (Regional Editor)

Professor (Retired) National Institute of Urban Affairs (NIUA) has a PhD in Environmental Science from JNU and an MPH from School of Public Health, University of California, Berkeley, United States. A national science talent scholar, a recipient of ICAR Fellowship in Agricultural Microbiology and Fogarty Int. fellowship in Environmental Health, she has presented in several conferences and seminars in India and abroad. She received the UNCHS Global 100 award for waste management and sanitation in Bangalore in 1998, Plasticon India award for Plastics Reuse and Recycling in 2005 and recognition from UNU-IAS, Japan for Regional Centre of Expertise on Pilgrimage Places in 2007.

Dr. Shyamala K. Mani joined NIUA in December 2012 after working at the Centre for Environment Education (CEE) for 25 years. A member of several professional organizations, she has published in reputed books and journals. She helped formulate the Biomedical Waste Management Rules 1998 and 2016, Municipal Solid

Waste Management (SWM) Rules 2000 and 2016, Plastics Waste Rules 2011 and 2016 promulgated by Ministry of Environment, and Forests and Climate Change. At NIUA, she has been involved in coordinating projects related to urban and regional sanitation, renewable energy planning, improving urban services through training and capacity building, and building climate change resilience in cities through education and innovation. She has been involved in capacity building and training of Urban Local Bodies (ULBs) under Swachh Bharat Mission (SBM) in 2016, 2017, and is currently Team Leader of the SBM SWM Exposure Workshops Project 2018–19. She is a member of the Expert Advisory Committee for selecting SWM projects for DST support, Chairperson of TIFAC committee for Technical Needs Assessment for the Waste sector for UNFCCC under Ministry of Science and Technology and a member of Integrated Solid Liquid Waste Management under Ministry of Drinking Water and Sanitation. She is also in the Advisory Committee for development of curriculum for IGNOU certificate course on Healthcare Waste Management, Solid and Liquid Waste Management, and Environmental Health & Safety.

Marina Yesica Recalde (Regional co-Editor) is an Argentinean economist specialized in Energy and Climate Change Economics. She holds a Ph.D. in Economics from Universidad Nacional del Sur. Currently she is researcher at the National Council of Science and Technology Argentina (CONICET) and the Department of Environment from Fundación Bariloche, Argentina.

Dr. Recalde made her PhD studies in Bahía Blanca (2005–2010), studying the link between energy resources, energy policies, and socioeconomic development in Argentina. In 2010, she made a short stay at the Universidad Autónoma de Barcelona, Spain. When she came back to Argentina, she worked as teacher assistant in Energy Economics. In 2012 she entered as a researcher in CONICET and in 2013 she started working as a (national and international) consultant in the field of energy and climate change.

As member of Fundación Bariloche, she has deeply and actively worked in the Climate Technology Centre & Network, in which Fundación Bariloche is one of the Consortium Members, as well as in the Technology Needs Assessment program form UNEP DTU.

Since 2013, she has provided technical assistance financed by different institutions (UNIDO, UNEP, BID, World Bank, and national governments from Latin American region, among others) in the area of energy policy, renewable energy, energy efficiency, M&E of energy efficiency, and technology transfer in Argentina, Colombia, Mexico, Paraguay, Uruguay, and other countries of the Latin American region. She has also worked as graduate and postgraduate teacher in the field of energy economics, energy regulation, and climate change in different public and private universities. She also has several papers and book chapters published in the branch of energy policies for renewable and energy efficiency promotion.

Dr. Atilio Armando Savino (Regional co-Editor)

Vice President of ARS-Association for Solid Waste Studies, National Member of ISWA in Argentina

Board Member and former President of ISWA—International Solid Waste Association

Chief Editor and author of the Waste Management Outlook for Latin America and the Caribbean—UNEP, 2018

Lead author of the Cross-cutting issue Waste in the Global Environment Outlook (GEO-6) assessment, UNEP, 2019

Former Secretary of Sanitary Health Determinants-Ministry of Health of Argentina

Former Secretary of Environment and Sustainable Development of Argentina-National Ministry of Health and Environment

Senior Consultant in Solid Waste Management

Vice President of COP11 Bureau (Conference of the Parties to the United Nations Framework Convention on Climate Change—COP UNFCCC) (2005–2006)

Atilio was born in Buenos Aires, Argentina on August 4, 1947. He is married with four children.

He obtained his title of Certified Public Accountant from the University of Buenos Aires, Argentina (UBA) in 1971.

In 1974 he got a Bachelor degree in Economics from the University of Buenos Aires, Argentina.

In 1985 he finished his Doctorate in Political Sciences at the University of Belgrano, Buenos Aires, Argentina, pending the presentation of the thesis.

At present, he is a senior consultant on solid waste management and climate change with an experience of more than 30 years.

He is now the Vice President of "Asociación para el Estudios de los Residuos Sólidos—ARS" (Association for Solid Waste Studies), Argentine National Member of the International Solid Waste Association—ISWA.

At the same time, he is a board member of the International Solid Waste Association—ISWA representing the Regional Developing Network of Latin America, and member of the International Advisory Board of ISWA's official journal "Waste Management & Research."

He was one of the participants at the preparation of the Global Waste Management Outlook (UNEP-ISWA 2015) and one of its reviewers.

He was the chief editor and author of the Waste Management Outlook for Latin America and the Caribbean—UNEP 2018, and lead-author of the cross-cutting issue "Waste" of the Geographical Environmental Outlook 6 (GEO 6)—UNEP 2019.

Between his different professional activities, it is worth mentioning that he was the General Manager of CEAMSE, a public company for the final disposal of waste of the Metropolitan Area of Buenos Aires, Argentina (16 million inhabitants) between 1992 and 2003.

From 2003 to 2006 he was the Secretary of Environment and Sustainable Development of Argentina, Vice President of COP 11 Bureau (Conference of the Parties to the United Nations Framework Convention on Climate Change—COP UNFCCC) in 2006 and Secretary of Sanitary Health Determinants, Ministry of Health of Argentina in 2007.

Between 2008 and 2010 he was the President of the International Solid Waste Association—ISWA.

Dr. William Stafford (Regional Editor) is a life scientist with 21 years of R&D covering topics ranging from biochemistry, microbial ecology, systems biology, bioenergy, permaculture, holistic resource management, industrial ecology, and sustainability science. Bioenergy and the bioeconomy is a current research focus which requires innovative solutions to meet development objectives of economic feasibility, social acceptance, and environmental protection. William has 26 publications in peer-reviewed scientific journals and is currently a researcher in the Green Economy Solutions competency area at the Council for Scientific and Industrial  Research (CSIR), and an extraordinary associate professor in the Department of Industrial Engineering, Stellenbosch University.

Foreword

In order to foster scientific research in the energy and sustainability sector, FAPESP launched a call for the creation of the Research Centre of Gas Innovation (RCGI). Its main mission is to be a world center for advanced studies in the sustainable use of natural gas, biogas, hydrogen, as well as the management, transportation, storage, and use of CO_2. The Centre, hosted at the University of São Paulo, is the result of FAPESP partnerships, in this case with SHELL Brasil, in support of high-level scientific research for the development of the energy sector. Its activities are based on three pillars: research, innovation, and dissemination of knowledge.

In this perspective, the book that is offered to the general public, and particularly to professionals and researchers in the renewable energy and sustainability sector, is fundamental because it presents the issue of the collection and disposal of solid waste in developing countries. The work in question has thematic pertinence to the subjects dealt with in the RCGI, from its main lines related to the action and promotion of energy sources with less impact to the environment, to the own sustainability in the production of biogas and energy conversion from waste.

The book clearly shows one of the key points for innovation in Brazil: the transformation of waste into energy (WtE) and its positive consequences for the environment. The costs for deploying WtE projects are discussed in the book, as well as a number of case studies, together with the challenges still existing in such countries.

The issue related to wealth generation and integration with energy production systems, in the particular case of WtE, their respective distribution and sustainability linked to the process are discussed in the book. This, without doubt, is already a success, since it has a function to awaken the reader to the problematic of applicability of finite resources in a sustainable way and with a view to reaching justice for present and future generations.

Julio R. Meneghini
Polytechnic School, RCGI, University of São Paulo, São Paulo, Brazil

Preface

The adequate collection and disposal of municipal solid waste (MSW) is a problem that has plagued urban centers since the antiquity. Even the city of Rome, the capital of the Roman Empire, more than 2000 years ago, did not have an adequate system of collection and disposal of urban waste despite having solved the problem of fresh water supply (with aqueducts) and disposal of liquid effluents (through the "maxima cloaca").

Currently the problem has been solved in urban centers of the industrialized countries in the Europe, North America, Japan, and a few others countries in Asia but it is still a major problem of municipalities in many developing countries.

This book discusses the reasons why they have not fully been solved yet. It gives special emphasis to waste to energy (WtE) production systems and the environmental consequences, which—for lack of information—have been an important hurdle to the widespread use of modern methods of disposing MSW.

The major problem of the high costs of the technologies needed is discussed as another important obstacle as well as the need for public policies to facilitate their adoption.

Case studies of several countries are presented.

The book serves as an excellent information source for government authorities and entrepreneurs concerned with such problems in many developing countries.

José Goldemberg
University of São Paulo, São Paulo, Brazil

Acknowledgments

The authors gratefully acknowledge the support from Shell Brazil and FAPESP through the "Research Centre for Gas Innovation—RCGI" (FAPESP Proc. 2014/50279-4), hosted at São Paulo University, and the strategic importance of the support given by ANP (Brazilian National Oil, Natural Gas and Biofuels Agency) through the R&D levy regulation. We also thank the Brazilian National Council for Scientific and Technological Development (CNPq) and Coordination for Improvement of Higher Education Personal (CAPES) for scholarships. Further acknowledgments must go to the Research Group on Bioenergy (GBIO), hosted at the Institute of Energy and Environment at the University of Sao Paulo.

Special acknowledgments to Professor José Goldemberg from University of São Paulo for the fruitful discussions and the Preface for the book.

We gratefully acknowledge the support of the R&D Project CESP-ANEEL 00061-0057/2017 "Electricity Cogeneration in the Sugar and Alcohol Sector Using Regional Bioenergetics: Technological Routes for Productive Process Optimization and Business Model for Generated Energy Commercialization."

Suani Teixeira Coelho

Full time Professor, Thesis Advisor, Graduate Program on Bioenergy, Institute of Energy and Environment, São Paulo, Brazil

Thesis Advisor, PhD Program on Bioenergy, USP/UNICAMP/UNESP, São Paulo, Brazil

Coordinator, Research Group on Bioenergy, Institute of Energy and Environment, São Paulo, Brazil

Deputy Coordinator, Economics and Policies Program, Research Center for Gas Innovation (RCGI/FAPESP/SHELL) University of São Paulo, São Paulo, Brazil

CHAPTER ONE

Introduction

Suani Teixeira Coelho*, Daniel Hugo Bouille[†], Shyamala K. Mani[‡],
William H.L. Stafford[§,¶]
*Research Group on Bioenergy, Institute of Energy and Environment, University of São Paulo, São Paulo, Brazil
[†]Fundación Bariloche, Río Negro, Argentina
[‡]National Institute of Urban Affairs (NIUA), India Habitat Centre, New Delhi, India
[§]Council for Scientific and Industrial Research, Stellenbosch, South Africa
[¶]Department of Industrial Engineering, University of Stellenbosch, Stellenbosch, South Africa

Biomass energy (e.g., bioenergy) can be produced from different feed stocks of biological origin, through several different processes to produce heat, electricity and transport fuels (i.e., biofuels). As stated in REN21 (2018), "bioenergy as solid fuels (biomass), liquids (biofuels), or gases (biogas or biomethane) can be used to produce heat for cooking and for space and water heating in the residential sector, in traditional stoves or in modern appliances such as pellet-fed central heating boilers." Besides that, bioenergy also can be used for cogeneration [combined production of electricity and heat—CHP (combined heat and power)]. One of the important sources of bioenergy corresponds to municipal wastes (solid waste and liquid effluents), which can be used for energy conversion, mainly electricity.

However, the adequate collection and disposal of MSW remain a challenge in DCs as a direct consequence of inadequate practices, which in turn produce negative environmental and social impacts.

In industrialized countries most MSW are collected, reused, recycled, and before being disposed in landfills, are recovered through WtE systems. The share of MSW in the biomass energy conversion worldwide is significant: in 2015, 18% of all biomass corresponded to MSW for heating and 4% to biogas from different sources. Considering electricity production, in 2015 biogas was responsible for 20% and MSW corresponded to 8% of electricity produced from biomass (REN21, 2018).

Fig. 1.1 illustrates the situation of MSW and WtE in European countries (Eurostat, 2018). Eurostat data shows that WtE (and recycling) is mostly used in the more developed European countries, landfills being used preferentially in the less industrialized European ones.

The situation is different in DCs, where the collection and adequate disposal of MSW are not yet a reality for most of their populations.

In Brazil, for instance, the adequate disposal of MSW is still a problem, especially in small and medium municipalities. There are more than 1000 municipalities generating about 42% of the total collected waste that have no adequate disposal in landfills. Moreover, there has been a significant increase in the specific waste generation (e.g., tones per

Municipal Solid Waste Energy Conversion in Developing Countries
https://doi.org/10.1016/B978-0-12-813419-1.00001-2

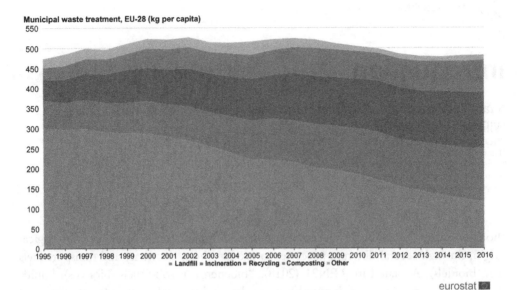

Fig. 1.1 Municipal waste treatment in the European Union from Eurostat. *(Data from Eurostat, 2018. Municipal Waste Treatment, EU-28 (kg per Capita).)*

capita per year) and adequate disposal does not follow this trend, largely in the North and Northeastern regions (ABRELPE, 2016).

In India, according to the Ministry of Housing and Urban Affairs (MoHUA), Government of India, the total generation of municipal waste is 145,128 metric tonnes per day (MT/D) of which 34.07% is processed (MoHUA, 2018). It also states that 79.5% of the wards in cities have achieved 100% door to door collection of domestic garbage. The current WtE production in India is 88.4 MW and the waste to compost production for September 2016 was 1506500 metric tonnes. Urban population in India is 37 million and the number of cities 4378 (MoHUA, 2018). In India state urban development departments and urban local bodies are responsible for waste management in the different cities. According to the Clean India campaign, which started on October 2, 2014 and will go on till October 1, 2019, all cities and towns will achieve 100% collection at doorstep and sizeable amount recovered and processed. According to the Solid Waste Management Rules 2016 promulgated by the Ministry of Environment, Forests and Climate Change, Government of India, not more than 10%–15% of the municipal waste should go for dumping, if at all only in sanitary landfills built according to specifications to prevent air, water and soil pollution. However, it is estimated that over 65% of the waste is currently going to dumpsites, which have no lining or preventive measures to stop air or water pollution.

In South Africa, a population of 52 million generates approximately 108 million tonnes of waste (Republic of South Africa, 2011). Municipalities are responsible for ensuring that adequate waste collection and disposal facilities are available. However, waste collection services are not fully rolled out, and almost one-third (30.1%) of households lack any kind of refuse facilities, particularly in small municipalities and rural areas, which are the most unserved areas. Waste disposal by landfill remains the most dominant method of disposal in South Africa. The reliance on landfill disposal, coupled with the relative low pricing for landfilling, has limited the incentive to devise alternative methods of dealing with waste. However, there are a few laudable examples where waste management practices have been recently improved with new measures and put in place to recover materials and energy from MSW. In addition, the recent development of national policy, namely the National Waste Management Strategy (NWMS), aims to drastically reduce waste to landfill through waste minimization, reuse, recycling and recovery of waste (DEA&DP, 2015) with a target of a 20% reduction in waste going to landfill by 2019.

Ahead, in this book, several other examples and difficulties are presented for African, Asian and Latin American countries. In addition, this book highlights that the main MSW disposal is still landfills and there are only a few examples on WtE processes. The main challenge is the economic feasibility since investment are extremely high and the demonstration plants remain very few. Besides these difficulties, there is another important challenge faced by DCs: energy access. As discussed in several publications, such as in BREA project (GBIO, 2015), access to cleaner and affordable energy options is essential for improving the livelihoods of the poor in DCs. In fact, there is a clear link between energy and poverty.

In DCs, there are 2.7 billion people (17% of world population) relying on traditional biomass for cooking and the overwhelming majority of the 1.2 billion have no access to electricity (38% of world population), despite some decrease in the energy deficit since 2010 (REN21, 2018). In addition, as shown in Fig. 1.2, a large percentage of such population lives in African and Asian countries.

For the least developed countries (LDC), the situation is much more dramatic, as shown in Fig. 1.3[1] (Traeger et al., 2017), but these countries are not the main objective of this study.

Coelho et al. (2015) and Coelho and Goldemberg (2013) discuss that energy access is still an important challenge faced by DCs (Coelho et al., 2015; Coelho and Goldemberg, 2013). The United Nations´ Secretary General Advisory Group, in 2010, defined it as one of the most important problems to be tackled in the next few decades (UN AGECC, 2010).

According to GEA (2012) and IIASA (2012), it is important the "access to affordable modern energy carriers and end-use conversion devices to improve living conditions and enhancing opportunities for economic development" (IIASA, 2012). In addition, here it is mentioned the idea of energy carriers for economic development instead of only "electricity access," reinforcing the idea of energy for productive uses.

---

[1] MSW collection and disposal, as well as energy access are huge challenges also for the LDCs, but in this text only the developing countries are analyzed.

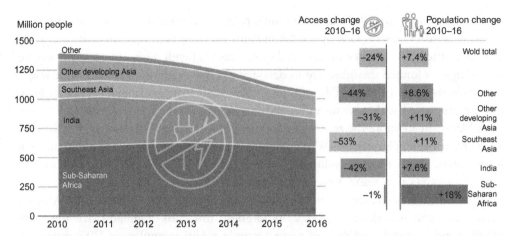

Fig. 1.2 Population without access to electricity, by region or country, 2010–16. *(Data from REN21, 2018. Global Status Report Renewables 2018 Global Status Report. Paris. http://www.ren21.net/gsr-2018/chapters/chapter_03/chapter_03/.)*

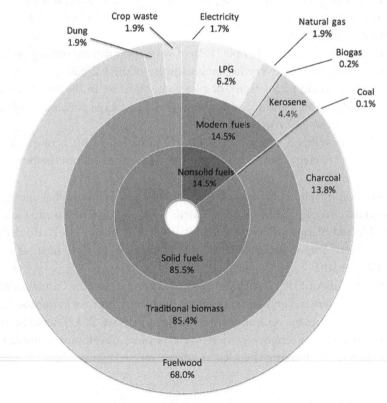

Fig. 1.3 Shares of different cooking fuel in LDCs. *(Data from Traeger, R., et al., 2017. ALDC: The Least Developed Countries Report 2017: Transformational Energy Access. Geneva.)*

As discussed by UN-Energy (2007), energy services are an essential input to economic development and social progress, notably for achieving the sustainable development goals. "Energy services are necessary for successful implementation of almost all sectorial development programs, notably revenue generating activities, health, education, water, food security, agricultural development, etc." Increased access to energy allows economic growth and poverty alleviation. The need for adequate policies to incentivize electricity access for economic development is stressed by UN-Energy (UN-Energy, 2011), which presents a set of recommendations aiming to accelerate sustainable electricity development programs for electricity access in DCs (on- and off-grid areas), identifying the most effective and meaningful best practices of partnerships. This study presents recommendations for adequate policies based on the idea that the improvement in electricity access needs a PPP—public-private partnership, considering the global goals of the AGECC report (UN AGECC, 2010).

As pointed out in BREA project (GBIO, 2015), the lack of modern and affordable forms of energy affects agricultural and economic productivity, opportunities for income generation, and more generally the ability to improve living conditions. Moreover, low agricultural and economic productivity, as well as diminished livelihood opportunities, in turn result in malnourishment, low earnings, and no or little surplus cash. This contributes to the poor remaining poor, and consequently they cannot afford to pay for cleaner or improved forms of energy such as fuels and equipment. In this sense, the problem of poverty remains closely intertwined with the lack of cleaner and affordable energy service.

It is well known that existing technologies commercially available include, as a first step, photovoltaic panels, but they are only able to generate energy for small types of power supply. In addition, they are quite expensive systems for DCs since most are still being imported. Therefore, the use of biomass residues allows the production of higher amounts of power, mainly in areas where such residues are available in large quantities.

The main advantage of the use of biomass residues is exactly the potential to produce power for productive activities, as well as contributing to reduce the negative impacts of the inadequate use of such residues (i.e., MSW and animal residues, which are often discharged inadequately in DCs).

Therefore, it is very important to analyze the synergies between MSW collection and adequate disposal, as well as the adoption of WtE technologies for energy production. The use of WtE for MSW corresponds to a win-win process, since it contributes to solve not only the problem of MSW disposal but also to increase the energy availability, which is particularly important issue in DCs facing difficulties for providing energy access.

There are several pathways for biomass conversion, depending upon the end use. Considering WtE conversion technologies, options are promising and almost all of them already being commercialized, as shown in Fig. 1.4, depending on the amount of waste, the local capacity building for operation & maintenance (O&M) and the funds and policies available.

It is important to note that WtE technologies using thermochemical routes are a controversial issue in some countries, since they face some negative perception from the local

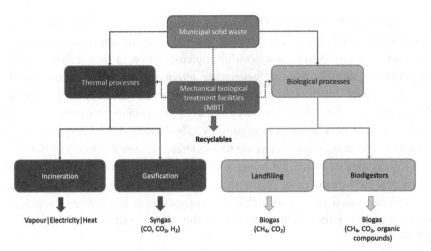

Fig. 1.4 WtE technological options (Coelho et al., 2018).

society. Among them, incineration process in many cases is the object of strong local rejection, for the fear of the pollutant emissions (mainly dioxins and furans). This has been, in a recent past, a challenge in European countries (such Portugal) and in DC´s such as Brazil. Personal visit[2] to Valorsul incineration plant in Lisbon allowed the understanding of such perception; local people feared the possible health impacts of emissions, due to the lack of information about the strict exhaust gas cleaning systems to comply with the European Union (EU) standards. In some DCs, a similar situation occurs, and it is worth to comment.

In Brazil, from 2002 to 2006, the state of Sao Paulo Environmental Secretariat developed a strong cooperation program with Bavarian Government (Germany) aiming to clarify these environmental aspects. During this period, the Secretariat—jointly with the Bavarian Government—developed a strong capacity-building program, not only for the state government technical agents, but also for the stakeholders involved. Later, in 2008, the State Environmental Agency published an important legislation establishing the limits for the pollutant emissions, based on the EU standards. This legislation is the basis for the environmental licensing of WtE plants in the state since then. However, there is only one incineration plant under construction in the country since investments are high and there is a huge difficulty to have enough funds for it.

Moreover, waste collection services remain a problem for small municipalities, since incineration plants are economically feasible only for municipalities with at least 500,000–600,000 inhabitants. In such cases, other options are necessary such as MSW gasification, which is possible to be installed in smaller municipalities.

---

[2] Personal communication, S. Coelho, GBIO/IEE/USP. 2011.

In this context, the main objective of this publication is to analyze the current situation of MSW collection and disposal, allowing the discussion of the perspectives of WtE in in Latin America, Africa, and Asia.

The DCs selected for discussion in this publication are the following:

- Latin America: Argentina, Brazil, Colombia, Ecuador, and Mexico.
- Asia: China, India, Indonesia, Japan[3], and Malaysia.
- Africa: Kenya, Malawi, Rwanda, South Africa, Uganda, and Zimbabwe.

In this context, Chapter 2 presents a general overview of the countries selected in each region, including socioeconomic figures such as energy access, share of urban and rural population, and environmental (i.e., environmental legislation and existing challenges) and social aspects [i.e., HDI (human development index), access to basic sanitation, waste collection and disposal].

Chapter 3 analyzes the best available technologies (BAT) for WtE conversion, both those already commercialized and those already available in each region. This chapter initially presents a general overview of WtE worldwide (biological and thermal treatments), as well as the general situation of the WtE market in the DCs of each region (local manufacturers, availability of the equipment locally, if imported or not).

Chapters 4–6 discuss the perspectives of the WtE technologies in the regions, how WtE can contribute to increase energy access, and present the selected case studies for each one of the regions (Latin America, Asia and Africa), focusing in the existing commercialized BATs.

Therefore, Chapters 4–6 discuss the main issues related to the subject at regional level:

- Existing MSW management and policies.
- Current situation of energy access and the existing difficulties to increase the access.
- Existing or suggested, the most adequate WtE.
- The WtE experiences (case studies).

Finally, Chapter 7 presents the main barriers found against WtE in the regions analyzed, and discusses the proposal of policies for the implementation of WtE technologies in countries, followed by the overall conclusions in Chapter 8.

---

[3] Despite being an industrialized country, Japan was included as a significant study case, with experiences to be followed.

CHAPTER TWO

Overview of Developing Countries

Suani Teixeira Coelho*, Marina Yesica Recalde†, Shyamala K. Mani‡,
William H.L. Stafford§,¶

*Research Group on Bioenergy, Institute of Energy and Environment, University of São Paulo, São Paulo, Brazil
†Fundación Bariloche, Buenos Aires, Argentina
‡National Institute of Urban Affairs (NIUA), India Habitat Centre, New Delhi, India
§Council for Scientific and Industrial Research, Stellenbosch, South Africa
¶Department of Industrial Engineering, University of Stellenbosch, Stellenbosch, South Africa

2.1 LATIN AMERICA: ECONOMIC, ENVIRONMENTAL, AND SOCIAL OVERVIEW

Marina Yesica Recalde
Fundación Bariloche, Buenos Aires, Argentina

The Human Development Index (HDI) can be used as an indicator to measure the socio-economic development, particularly, considering the links between waste, energy, and other aspects. As defined by UNEP (United Nations Environment Programme), the HDI is a composite index measuring average achievement in three basic dimensions of human development: a long and healthy life, knowledge, and a decent standard of living. Also in other socioeconomic indicators, the HDI of LAC (Latin American Countries) is quite good in comparison to other developing regions of the world, occupying in 2015 the second place for the developing regions (0.751), after the East Asia and the Pacific (0.720) and far away from sub-Saharan Africa (0.523). As this index is clearly country specific and the situation inside the LAC is very dissimilar it is good to have an idea on the situation of the countries under analysis in this book regarding the rank. In this sense, in 2015 Argentina ranked 45 (0.827), México 77 (0.762), Brazil 79 (0.754), Ecuador 89 (0.739), and Colombia 95 (0.727).

The gross national income (GNI) per capita (2011 PPP USD) in 2015 for the LAC was on average 14,028, while for the rest of the developing regions the GNI per capita was between 14,958 (Arab States) and 3383 (Sub-Saharan Africa). However, this value is far away from the most developed regions: in the group of very high human development countries, the GNI per capita was 39,605. Most of the countries selected for the case studies display an income inequality distribution as shown by the Gini Index, which is above 40 for all of them (above 50 in the case of Brazil and Colombia). Conversely, in most of OECD (Organization for Economic Co-operation and Development) countries this

Municipal Solid Waste Energy Conversion in Developing Countries
https://doi.org/10.1016/B978-0-12-813419-1.00002-4

index is around 31, differing significantly among countries, for instance, Germany (31.7), Austria (30.5), Belgium (27.7), Denmark (28.2), Czech Republic (25.69), France (32.7), Hungary (30.4), Spain (36.2), Italy (34.7), and United States (41.5), among others.

Except for Colombia and Ecuador, the countries under analysis have high urbanization rates and good electrification rates (at least in urban areas), which, according to the World Bank Sustainable Energy for All (SE4ALL) database, is nearly 99/100% in all the cases for total electricity access.[1] Concerning electricity consumption, electricity per capita in the LAC region is also superior to other underdeveloped or developing regions of the world (2.1 MWh/capita in 2015), but significantly lower than OECD countries (8.02 MWh/capita in 2015) (Table 2.1).

In the last few years, the debate on GHG emissions in the energy sector and its environmental impact has become more relevant for the developing regions. This situation is expected to deepen in the following years as emissions have grown more significantly in nondeveloped world in relative terms, directly related to GHG emissions in the energy sector and its key drivers: demographic trends, economic activity and income, and technological and structural changes (Recalde et al., 2014). Indeed, energy-related emission grew significantly in developing non-OECD countries in the period 2001–12 (nearly 80%) and remained nearly stable in the OECD region (DOI/IEA, 2013). Additionally, different policy scenarios stress that developing countries will be the most energy consuming and emitting countries, because their economies and population will grow at a higher rate than developed ones, and will probably "rely on fossil fuels to meet this fast-paced growth in energy demand" (DOI/IEA, 2013; IPCC, 2013). This will increase the relevance in the developing countries for the implementation of mitigation policies.

Table 2.1 Country Socio-Economic and Energy Specific Characteristics

Country	HDI Rank[a]	GNI pc[2015] (2011 PPP USD)	GINI Index[b]	Urban Population (% of Total)[c]	Elec. Cons. pc (MWh/ capita)[d]	CO2/Pop (tCO2/ capita)[d]	CO2/GDP (PPP) (kgCO2/2010 USD)[d]	(TPES)/GDP PPP (toe/mil 2010 USD)[d]
Argentina	45	20,945	42.4 (2016)	92	3.09	4.41	4.41	0.12
Brazil	79	14,145	51.3 (2015)	86	2.52	2.17	0.15	0.1
Colombia	95	12,762	50.8 (2016)	77	1.23	1.5	0.12	0.06
Ecuador	89	10,536	45 (2016)	64	1.43	2.33	0.22	0.09
Mexico	77	16,383	43.4 (2016)	80	2.23	3.66	0.22	0.09

[a]Base 2015. Source: UNDP-HDI (2016).
[b]Base 2016, except for Brazil: 2015. Values between 0 and 100. Source: World DataBank: http://databank.worldbank.org.
[c]Source: World Bank database, based on United Nations Population Division. World Urbanization Prospects: 2014 Revision.
[d]Base 2015, Source: IEA Key World Energy Statistics 2015: www.iea.org.
Source: Adapted from Recalde, M.Y., 2016. The different paths for renewable energies in Latin American Countries: the relevance of the enabling frameworks and the design of instruments. WIREs Energy Environ. 5(3), 305–326.

[1] https://data.worldbank.org/indicator/EG.ELC.ACCS.ZS.

Environmental impact, especially in terms of climate change impact, can be measured differently. This may have a different impact in terms of the climate change regional responsibilities debate: total CO_2 emissions (1132.47 $MtCO_2$ in LAC and 11,720.23 $MtCO_2$ in OECD), CO_2 emissions by GDP (0.18 and 0.25 $kgCO_2$/2010 USD in LAC and OECD respectively), or as CO_2 emissions per capita (2.33 and 9.18 tCO_2/population in LAC and OECD, respectively). It is straightforward that the environmental impact of the region (measured by these indicators) is very low compared to the majority of the regions of the world (OECD, Middle East, Non-OECD Europe and Eurasia, and China). However, this may not the case in some specific countries of the regions (e.g., Argentina and México), which can be, to some extent, related to their electricity generation mix.

The LAC presents one of the cleanest electricity mix of the world, with more than 50% corresponding to big hydro and nearly 42% to thermal and nuclear power generation (the remaining share corresponds to nonconventional renewable energy sources (NRES)).

The situation within the region is diverse; it can be divided into countries where big hydropower generation is predominant (Paraguay, Guatemala, Colombia, Costa Rica, Brazil, Suriname, Venezuela, Belize, Uruguay, Peru, and Ecuador) and countries with higher share of thermal power generation (Panamá, Bolivia, Argentina, Chile, Haiti, México, Dominican Republic, Jamaica, Cuba, Guyana, Barbados, and Grenada) (Recalde, 2016). However, in the past decades many of these countries have been facing different energy challenges: increasing energy security, reducing external dependence and economic impact of the energy balance, and improving environmental quality of the energy sector, among others. This led them to enhance their efforts to implement the energy policies to promote the diversification of their electricity sectors. Therefore, they have begun the path to the promotion of renewable energies in electricity sectors, and the use of waste-to-energy (WtE) technologies. Therefore, as it will be discussed in this chapter and Chapter 4, the LAC region, and the countries selected as a case of study have begun to promote deeply the use of other energy sources in their energy systems, both as a way to increase energy access and reduce energy environmental impact.

Finally, and directly related to the climate change debate and impact, all the 33 Latin American nations belong to the UNFCCC, all of them (except for Nicaragua) also belong to the Paris Accord and 31 countries have already presented their NDCs (nationally determined contributions). All the NDCs presented by the countries have conditional and unconditional targets, and many of them include the energy sector (and some mention the waste sector) among the most relevant sectors for mitigation policies (UNEP, 2016).

Section 2.1 presents a brief characterization of the economic, social, and environmental situation of each one of the five LACs analyzed in the book.

2.1.1 Economic Overview

2.1.1.1 Argentina—Economic Overview

Alejandro Cittadino* and Atilio Armando Savino[†]

*Departamento de Ecología, Genética y Evolución—Facultad de Ciencias Exactas y Naturales, Universidad de Buenos Aires, Buenos Aires, Argentina
[†]International Solid Waste Association (ISWA), Buenos Aires, Argentina

The Argentine Republic, located in the southernmost part of the American Continent, has a surface of $3,761,274 \, km^2$, of which $2,780,400 \, km^2$ are continental and the rest corresponds to the Antarctic area and the South Atlantic islands. The administrative division of the Argentine Republic includes 23 provinces and the Autonomous City of Buenos Aires (CABA).

Argentina is one of the largest economies in the Latin American region, with a GDP (constant 2010 U$S) of 460,334 billion U$S in 2017. The country is characterized by a macroeconomic volatility, with periods of acceleration followed by recessions or deep crisis. In the last two decades the Argentinean GDP grew steadily after the 2001 economic crisis (258,282 billion 2010 US$) until 2008 (408,877 billion 2010 US$), mainly as the result of a combination of aspects related to the international crises and some internal factors. Then, between 2011 and 2017 the country had minor volatilities (Fig. 2.1).

The country is very well dotted of natural resources, particularly for agricultural production and energy (nonrenewable and renewable), which has made it one of the largest food producers in the world with large-scale agricultural and livestock industries. Argentinean economic output is based largely on industrial production and an export-oriented agricultural sector, which account for 63% of total exports. However, as stated by PNUD (2017), since the mid-1970s Argentinean economy has been subjected to endogenous and exogenous impacts resulting in a deindustrialization process, labor precarity, and long-term economic growth below the potential.

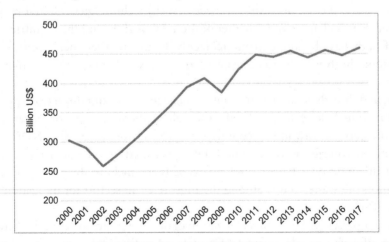

Fig. 2.1 Gross domestic product (GDP) (constant 2010 US$) in Argentina *(Source: World Bank database.)*

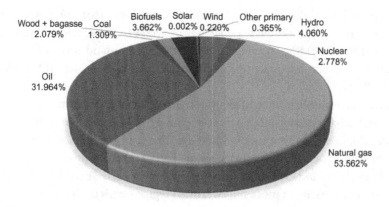

Fig. 2.2 Total primary energy sources in Argentina in 2016.

Despite its high renewable energy endowment and the existence of policies to promote the use of renewable energies (Recalde, 2016a,b), Argentina is still highly dependent on hydrocarbons. According to the National Energy Balance (BEN), published by the Ministry of Energy (MEN) in 2016, the total primary energy supply (TPES) was 80,060 tons of oil equivalent (toe), with natural gas and oil accounting for 86% of total TPES, as shown in Fig. 2.2.

2.1.1.2 Brazil—Economic Overview

Luís Gustavo Tudeschini
Research Group on Bioenergy, Institute of Energy and Environment, University of São Paulo, São Paulo, Brazil

The Brazilian population is projected to grow until 2042 and reach a total of 228 mi inhabitants (IBGE, 2013). If the current production and consumption behavior maintains, it would generate around 237,120 tons of municipal solid waste (MSW) per day and demand 282 mi toe of energy per year.[2] The adoption of waste-to-energy (WtE) technologies gives a sustainable path to attend both the demand for a sustainable MSW management (MSWM) and power generation.

The National Plan on Solid Waste[3] (NPSW), approved in August 2010, is the country's central regulatory framework on solid waste (SW) management, and its fundamental objectives promote healthier and environmentally friendly pathways to manage solid in the 5570 municipalities. Nevertheless, the MSWM in Brazil shows slight improvement and still needs to address essential challenges. Those challenges consist of providing waste collection, adequate processing and disposal, reusing and recycling and, most especially, explore the potential for energy generation.

---

[2] Using reference values from 2016. Municipal solid waste generation: 1.04 kg/capita/day (ABRELPE, 2016), and final energy demand: 1.24 toe/capita/year (MME/EPE, 2017).

[3] Federal Law number 12.305, August 2, 2010 (Brasil, 2010).

From 2010 to 2016,[4] the data on SW management shows that: the population share with access to waste collection decreased 7% (from 82% to 75%); and adequate disposal increased less than 1% (0.8%), leaving almost 40% (81 thousand tons) of the collected waste in 2016 without adequate disposal (ABRELPE, 2016; MCIDADES.SNSA, 2012; SNIS, 2016).

These are average numbers often mask important regional particularities. Each of the five geographic regions has extreme differences concerning the physical environment and socio-economic development. As an example, the GDP per capita in the Southeast region, the richest and most populous, is more than two times higher when compared with the Northeast region, US$11,109 and US$ 4412[5] per year respectively (IBGE, 2017). Consequently, the low investment capacity in more impoverished regions leads to the worst overall waste management indicators. As shown in Table 2.2, when compared with the national level, the North and Northeast regions have the lowest economic performance, which is translated into worse MSWM. Within regions, these discrepancies are even more considerable. For instance, in Acre (a northern state) 45% of the population has not access to waste collection, while in the state of Amapa (also a northern state) this number is 21% lower.

The cost of the current system also varies geographically, especially at the municipal level. The national average MSWM cost is US$ 31.80 per capita, but half of the cities spend less than US$26.00 for these services. On the other hand, there are extreme cases as in the insular municipality of Fernando de Noronha, which the cost of MSWM is about US$802.00 per capita. Thus, the share of these costs in the budget presents significant variations being the national average 3.4%, but more than 100 municipalities spend over 10% of its funds on these services.

Moreover, the municipalities in less developed areas also lack access to modern and affordable energy carriers. With the objective of overcoming this issue, an important step

Table 2.2 Regional Differences in WM and Economic Indicators

	Pop. With Access to Waste Collection (%)	Urban Pop. With Selective Waste Collection (%)	Share of MWM Cost on Municipal Budget (%)	MWM Cost (US$[a]/ capita)	GDP per Capita 2015 (US$[a]/year)
North	93.1	26.0	3.7	25.4	5400
Northeast	94.7	40.3	4.1	33.4	4412
Southeast	98.7	67.4	3.4	29.6	11,109
South	97.9	83.5	2.7	34.7	10,143
Midwest	98.1	67.6	3.6	32.0	11,038
Brazil	97.1	73.3	3.4	31.8	8491

[a]Conversion rate: 3.40 BRL/USD.
Source: SNIS, 2016. Sistema Nacional de Informações Sobre Saneamento. http://app.cidades.gov.br/serieHistorica/; IBGE, 2017. Contas Regionais 2015: Queda No PIB Atinge Todas as Unidades Da Federação Pela Primeira Vez Na Série.

[4] Most recent available data (ABRELPE, 2016).
[5] Using conversion rate of 3.40 BRL/USD.

was taken with the Light for All[6] program, which focused on providing energy for basic needs and allowed access to more than 99% of the urban population. A further step to take is supplying energy for productive activities (i.e., water pumping, irrigation, and agricultural processes), promoting economic development, generating income, and improving the living conditions (Coelho et al., 2015; Coelho and Goldemberg, 2013). The adoption of WtE systems is an important option to help implement the second step since it has the potential to produce higher amounts of energy for productive activities (Coelho et al., 2015; Sanches-Pereira et al., 2016).

In conclusion, although the NPSW is a vital starting point in dealing with the waste management problems and providing a holistic regulatory framework, it presents poor results, especially in less developed municipalities. The use of WtE systems in the municipal SWM offers the opportunity to deal with these issues once this synergy can help expand the energy supply and contribute to reducing the negative impacts of the inadequate disposal. Albeit, it will require long-term planning and investment, besides the appropriate regulatory framework that considers the regional disparities.

2.1.1.3 Ecuador—Economic Overview

Rafael Soria and Laura Salgado

Departamento de Ingeniería Mecánica, Escuela Politécnica Nacional, Ladrón de Guevara, Quito, Ecuador

According to the World Bank, the total GDP (current prices) has grown significantly from 2000 (18.32 billion US$) until 2017, when Ecuador registered the highest value of its history with 103.06 billion US$ (World Bank, 2017). The GDP per capita (at current prices) in 2017 was 6199 US$ (World Bank, 2017). On the other hand, according to the Central Bank of Ecuador (BCE), the annual GDP growth rate in 2017 was 0.7%, which for 2017 would conduct a total GDP of 100.5 billion US$ (this is a prevision, so far there is no official value) (BCE, 2017a).

Historically in Ecuador, the oil resources have been the key driver for economic development. According to the Ministry of Hydrocarbons of Ecuador, the proven oil reserves (1P) of Ecuador by the end of 2017 were 1703 million barrels (Ministerio de Hidrocarburos, 2018). During the last decade Ecuador has produced an average of 519,000 oil barrels/day and has exported an average of 349,000 oil barrels/day (MICSE, 2016; OLADE, 2017). Based on that information, the indicator of horizon of proven reserves, the proved reserves/annual average production (R/P) would be 9 years. This indicator is an alert to start planning the medium and long-term energy system and an economic transition in Ecuador. One of the options to supply additional energy is the energetic use of waste (WtE) by using MSW, agro-industrial residues, and sewage water as sources of energy.

---

[6] *Luz Para Todos*, in Portuguese.

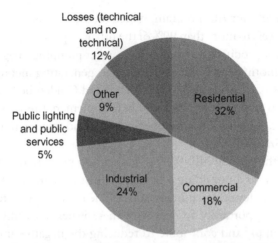

Fig. 2.3 Public service electricity consumption by sector. Total: 22 TWh in 2016. *(Source: Own elaboration, based on MEER (2017).)*

In January 2017 oil exports (568 million US$) represented 36% of total exports (BCE, 2017b). Nonoil exports are based on shrimps, banana, cacao, tuna & fish, and coffee. Imports are mostly associated with raw materials, capital goods, consumption goods, and oil products. In 2017 Ecuador had a positive commercial balance of 91.53 million US$ (including 3.71 billion of oil trades and 3.62 billion of nonoil trades) (BCE, 2018).

Ecuador's 2015 total energy consumption, 94.68 millions of barrels of oil equivalent (BOE) is characterized by the consumption of fossil fuels (78%), followed by electricity (15%), and the remaining by other minor sources (firewood and biomass) (MICSE, 2016). Total final energy consumption was 89.32 million BOE. The transport sector is the largest final energy consumer (46%) (gasoline and diesel), followed by industry (19%), and the residential sector (13%), with the remaining energy used by the commercial and other sectors. Total annual electricity demand has grown at an average rate of 5.8%/year over the last decade (2007–16) (MEER, 2017). The total electricity consumption reached 22 TWh in 2016; the electricity consumption share by sector is presented in Fig. 2.3.

2.1.1.4 Mexico—Economic Overview

Gustavo Solórzano
Asociación Mexicana de Ingeniería, Ciencia y Gestión Ambiental, A.C. (AMICA), Mexico City, Mexico

Mexico is located in North America, with a surface area of 1,964,375 km²; it is the world's 13th largest country by total area, and the 5th largest in the Americas. Mexico's total population reached 123.5 million inhabitants in 2017; Mexico City, the country's capital city, is the most densely populated area with 5967 inhabitants/km², while the national average was 61 inhabitants/km² in 2015 (CONAPO, 2017).

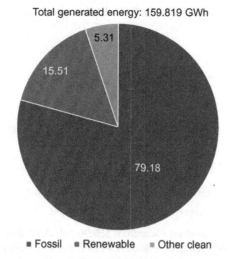

Total generated energy: 159.819 GWh

■ Fossil   ■ Renewable   ■ Other clean

Fig. 2.4 Power generation sources (first half 2017). *(Data from SENER, S.d., 2017. Reporte de avance de energías limpias. 1er semestre 2017. SENER, Ciudad de México. Retrieved April 23, 2018.)*

According to the World Bank, Mexico reported a GDP per capita (current USD) equivalent to 8902.8 in 2017, while its GDP (current USD) for the same year was 1.285 trillion US$ (constant 2010 US$), positioning Mexico as the second largest economy among countries in the LAC region.

Since 2004, the mean annual growth rate for energy generation with renewable sources has been 3.2%. By the end of first half of 2017, Mexico generated 20.82% of electric power from clean sources. As of June 30, 2017, clean energy-based power generation reached 33,274.31 GWh, which is equivalent to 8.79% increase compared to the first half of 2016. Fig. 2.4 illustrates power generation sources (first half of 2017) and show energy sources, total and renewable. It is important to note that although Mexico is an oil-producing country, it is committed to increase the contribution of clean energies to power generation to 35% of the share by 2024 (SENER, 2017) (Fig. 2.5).

2.1.2 Environmental Overview
2.1.2.1 Argentina—Environmental Overview

Estela Santalla

Departamento Ingeniería Química, Facultad de Ingeniería/UNICEN, Buenos Aires, Argentina

According to the Second Biennial Update Report of Argentina to the UNFCCC (2017), the waste sector generated 13.9 $MtCO_2e$ in 2014 being MSW disposed on land responsible for 49% of these emissions. From the total greenhouse gas (GHG) emissions of Argentina, MSW represents 2% (6.81 $MtCO_2e$) while industrial wastewater management (3.38 $MtCO_2e$), domestic wastewater (3.69 $MtCO_2e$), and the manure management from agricultural sector (2.14 $MtCO_2e$) each one represent approximately 1%. Fig. 2.6 details the Argentinean GHG emissions pathway in the different scenarios evaluated.

Total 33.274 GWh (15.51% of all types)

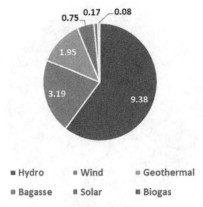

Fig. 2.5 Power from renewable sources. *(Data from SENER, S.d., 2017. Reporte de avance de energías limpias. 1er semestre 2017. SENER, Ciudad de México. Retrieved April 23, 2018.)*

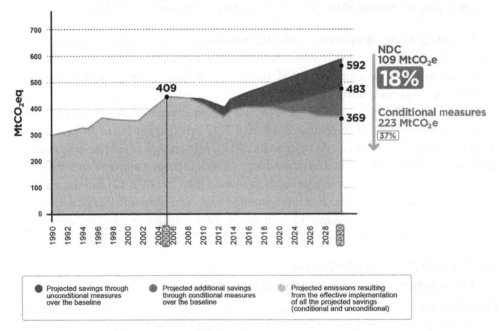

Fig. 2.6 Emissions pathway in the BAU, unconditional and conditioned measures scenarios. *(Source: http://www4.unfccc.int/ndcregistry/PublishedDocuments/Argentina%20First/Traducci%C3%B3n% 20NDC_Argentina.pdf.)*

According to the Decisions 1/CP.19 and 1/CP.20 of the United Nations Framework Convention on Climate Change, Argentina presented on October 1, 2015 its Intended nationally determined Contributions (INDC), which was revised after the Paris Agreement. The NDC presented a new goal of CO_2e emissions as result of the mitigation

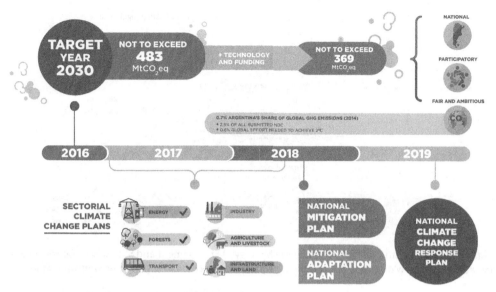

Fig. 2.7 Nationally determined contribution of Argentina. *(Source: https://www.argentina.gob.ar/ambiente/sustentabilidad/planes-sectoriales.)*

measures proposed for 2030 in energy, forest, transport, industry, agriculture-livestock, infrastructure, and land sectors. These goals were performed through the corresponding sectoral climate change plans (Fig. 2.7), which for the waste sector, for 2018, is not still elaborated.

The Ministry of Environment and Sustainable Development, through the Secretary of Control and Environmental Monitoring,[7] as the result of the 2020 Implementation Tables *Recycling of different waste streams*, participated by most of the actors of the civil society linked to the waste sector, elaborated a document that assumes the current regulations regarding waste management. The purpose is to build, on its base, the desirable normative structure in matter of waste, in light of the identified needs. Such information indicates that the current regulation poses a series of difficulties that hinder the implementation of public policies according to the current context and management needs. Among the most significant points, which can alter the current scenario of technological decisions for the treatment of the MSW, is the following are proposed:

- Decrease the generation of waste and promote its proper management through the establishment of extended producer responsibility.
- Collect the guiding principles in waste management such as gradualism, extended responsibility, shared responsibility, integrated life cycle, from cradle to cradle, and from cradle to grave.
- Establish the responsibilities of the subjects that intervene during the life cycle (generator, transporter, and operator), best available practices, traceability.

---

[7] https://www.argentina.gob.ar/ambiente/preservacion-control/estructura-normativa-residuos.

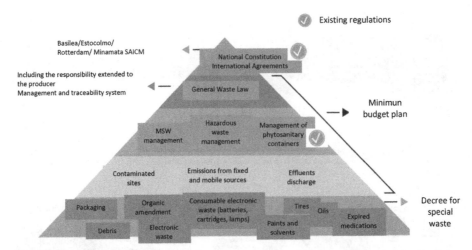

Fig. 2.8 Normative structure proposed by the Ministry of Environment and Sustainable Development. *(Source: https://www.argentina.gob.ar/ambiente/preservacion-control/estructura-normativa-residuos.)*

The document also claims for a law of minimum budgets that allows the updating of the technical criteria, delimits competences and criteria for the control of the atmospheric emissions, contemplating a traceability system for the monitoring and control of the waste, unifying the management at the national level national, and creating the corresponding records. In addition, it should contemplate a mandatory contribution for the management of each product placed on the market when it is finally disposed by the consumer. Fig. 2.8 summarizes the proposal of the new normative structure for the waste sector.

Within the clean development mechanism (CDM) scenario, five projects of GHG mitigation based on biodigestion technology were registered in Waste Handling and Disposal category. All of them belonged to the agro-industrial sector and focused on the capture of methane from wastewater and the use for thermal or fossil fuel replacement; nevertheless, none of these projects achieved the issuance of CERs (Blanco et al., 2016).

A study on the environmental impacts of the intensive production of milk (dairy) and meat (beef, pig, and poultry) in Argentina revealed that these activities consider neither the final destination of the waste nor the lifecycle of the product. Thus, in the usual practice treatments of wastewater or manure until discharge to soil or to the nearest surface water course are almost absent, ignoring the buffering capacity of the ecosystem to absorb them (Córdoba et al., 2008). The significant GHG mitigation potential of these sectors was confirmed through another study based on a resource assessment of carbon offsets in the most productive regions of the country, which revealed that sugar distilleries resulted the sector with the highest potential followed by swine, dairy, slaughterhouses, and citrus processing (USEPA, 2009).

A preliminary assessment of technologies for GHG mitigation in the waste sector in Argentina revealed that the capture of methane and fossil fuel substitution for energy use (thermal and/or electricity) by applying biodigestion in the agro-industrial sector (slaughterhouse, citric, sugar, and dairy) could achieve a reduction of approximately 700,000 tCO_2eq per year (TNA, 2012).

2.1.2.2 Brazil—Environmental Overview

Vanessa Pecora Garcilasso and Fernando C. de Oliveira
Research Group on Bioenergy, Institute of Energy and Environment, University of São Paulo, São Paulo, Brazil

In 2015, Brazil has emitted around 64 million tons of carbon dioxide related to waste sector, which represent around 3.34% of national emissions. In the period from 1970 to 2015, the cumulative volume corresponds to 1582 million tons of CO_2e or 2.38% of the total accumulated emissions in Brazil. Despite its low contribution, the waste sector is marked by strong growth (ICLEI, 2017).

Proper waste handling is an important strategy for environmental preservation and for the promotion and protection of health. Once packed in controlled landfills or dumps, SWs may compromise the quality of soil, water, and air, as they are sources of volatile organic compounds, pesticides, solvents, and heavy metals (DEWHA, 2009).

The decomposition of the organic matter present in the waste results in the formation of a dark-colored liquid, called slurry, which can contaminate the soil and the surface water by contaminating the groundwater. There may also be the formation of toxic gases, asphyxiation, and explosives that accumulate underground or are thrown in the air (Gouveia et al., 2010).

The storage and final disposal sites become favorable environments for the proliferation of vectors and other disease-transmitting agents. There may also be the emission of particles and air pollutants directly from the burning of garbage outdoors or incineration of waste without the use of appropriate control equipment. In general, these degradation impacts extend beyond the areas of final waste disposal, affecting the entire population.

There are also risks to people's health, especially for the professionals most directly involved with the handling of the waste, as in the case of operational staff of the sector, which mostly does not have the least measures of occupational prevention and safety. The situation becomes even more critical for individuals who work and live in the recovery of waste materials, who perform their work in very unhealthy conditions, usually without protective equipment, resulting in high likelihood of acquiring diseases. These people, who work directly with waste, especially the pickers of recyclable materials, are specifically vulnerable to these issues.

The treatment and disposal of MSW also encompass a set of activities and processes that generate direct and indirect jobs. Such activities range from collecting, sorting, and recycling that involve several processes of transformation of materials into new products, treatment technologies, energy recovery, and the final disposal (DEWHA, 2009).

2.1.2.3 Ecuador—Environmental Overview

Rafael Soria and Laura Salgado
Departamento de Ingeniería Mecánica, Escuela Politécnica Nacional, Ladrón de Guevara, Quito, Ecuador

Although Ecuador generates less than 0.5% of the global GHG emissions causing global climate change, it is voluntarily committed to face this challenge. In this way, in 1992 Ecuador signed the UNFCCC; in 1999, the Kyoto Protocol; in 2016, the Paris Agreement and, consequently, Ecuador participates of international negotiations about climate change and locally generates the regulatory and institutional framework to comply with the objectives set by the UNFCCC.

Ecuador presented to UNFCCC three National Communications about Climate Change (in 2001, 2011, and 2017). The Third National Communication updated the inventory of GHG for 2012, and reported the efforts and achievements of Ecuador in terms of mitigation and adaptation to climate change in the period 2011–15 (MAE, 2017a).

Fig. 2.9 presents the most updated official data, from the Third National Communication, about the evolution of GHG emissions, from 85 million tCO_2e in 1994 to 81 million tCO_2e in 2012. Most of the emissions were generated by the energy sector (46.6%), followed by net LULUCF (land use, land-use change, and forestry) emissions

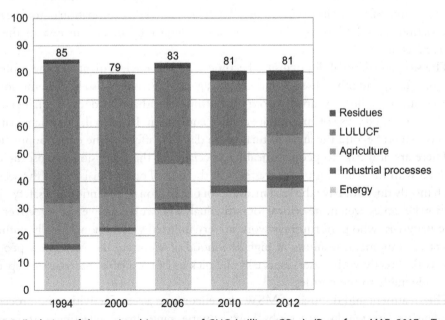

Fig. 2.9 Evolution of the national inventory of GHG (million tCO_2e). *(Data from MAE, 2017a. Tercera Comunicación Nacional del Ecuador a la Convención Marco de las Naciones Unidas sobre el Cambio Climático, first ed. MAE, Quito, Ecuador.)*

(25.3%). Third place is occupied by the agriculture sector, with 18.2% of GHG emitted to the atmosphere. The industrial processes and residues sectors represent, in total, approximately 10% of the Ecuadorian emissions. The emissions generated by the residues sector were 3.4 million tCO_2e in 2012, 87% by final disposal activities of solid residues and 17% by sewage water disposal. The GHG emissions of the residues sector represents 4.2% of total national GHG emissions (MAE, 2017a).

LULUCF emissions showed a sustained reduction in net GHG emissions throughout the period 1994–2012, mainly due to the increment in absorptions and the reduction in emissions in the "Land converted to agricultural land" category. The GHG emissions in the energy sector grew from 15 million tCO_2e in 1994 to 38 million tCO_2e in 2012. Within the energy sector, only 45% of the emissions are related to fuel combustion in transport activities, showing the importance of fostering mitigation activities in this subsector.

Aware of the adverse effects of climate change and in strict respect to international agreements, Ecuador signed the Paris Agreement on climate change. The Intended Ecuadorian Nationally Determined Contribution were submitted to COP21 on October 13, 2015 (UNFCCC, 2015).

Table 2.3 summarizes the INDC of Ecuador for 2025. These commitments did not consider any specific INDC in the residues sector. The first NDC was prepared and presented in March 2019, which also incorporated specific commitments for the residues sector.

2.1.2.4 Mexico—Environmental Overview

Gustavo Solórzano

Asociación Mexicana de Ingeniería, Ciencia y Gestión Ambiental, A.C. (AMICA), Mexico City, Mexico

Mexico's GHG emissions reached 683 Mt of CO_2e in 2015 excluding LULUCF, or 535 Mt of CO_2e if this last category is considered. Methane, an important GHG related to waste sector, represented 21% of the total emissions, equivalent to nearly 144 $MtCO_2e$.

The GHG emissions from waste sector (INECC, 2018, as shown in Fig. 2.10) represented 7% in 2015, equivalent to 46 Mt of CO_2e for the same year (contribution of each waste category to this figure is shown in INECC, 2018) (Fig. 2.11), including wastewater treatment. This reveals an important increasing trend for the sector, as in 2013 emissions from the waste sector reached only 31 Mt of CO_2e, equivalent to 4.6% of GHG total emissions (665 Mt CO_2e) (INECC, 2018). Regarding contributions by gas type, it is interesting to note that in 2015 nearly 94% of the GHG emissions from the waste sector corresponded to methane.

Mexico has committed unconditionally to carry out mitigation actions that would reduce its GHG emissions by 22% by 2030, which is equivalent to a reduction of 210 Mt of total CO_2e emissions, with 35 Mt corresponding to the waste sector.

Table 2.3 Ecuador's INDCs Commitments for 2025

Sector	Subsector	INDC—Unconditional (Additional to the BAU)	INDC Conditioned to International Support (Additional to the BAU)
INDCs–Energy	Power	Deployment of 2828 MW of hydroelectric plants up to 2025	Deployment of 4382 MW of hydroelectric plants up to 2025
	Oil and gas, power	Optimization of the use of associated gas from oil exploitation in the Amazon region to produce electricity. This electricity would replace diesel and would be use to supply the power demand from oil industry, water pumping, oil industry camps and communities in covered areas (sites closed to oil fields)	NA
	Oil and gas, power	Link of isolated power systems from oil industry in the Amazon region with the National Interconnected Power System (SNI)	NA
	Residential sector	The incorporation of 1,500,000 induction stoves, replacing LPG stoves	The incorporation of 4,300,000 induction stoves, replacing LPG stoves
INDC-LULUCF	Forestry and management of protected areas	Through the National Forestry Restoration Program, Ecuador plans to restore 500,000 additional hectares (in comparison to the restored areas until 2015) until 2017 and increase this total by 100,000 ha per year until 2025	Adding an additional 2 million hectares of restored areas in 2017

Source: Own elaboration, based on UNFCCC (2015).

Mexico has also committed to achieving a reduction of 51% in its black carbon emissions by the same year (SEMARNAT, 2015a,b).

In 2015, a total of 54.1 million tons of MSW were produced in Mexico. This figure represents an increase of 61.2% when compared to 10.24 million tons of similar waste generated in 2003. The per capita generation reached 1.2 kg/day in 2015, with higher values in large cities and in cities along the border with the United States (SEMARNAT, 2016).

Total emissions: 683 Mt CO$_2$e

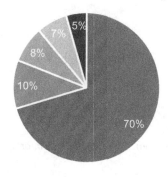

■ Energy   ■ Cattle   ■ Industry   ■ Waste   ■ Aggregated sources

Fig. 2.10 GHG emissions by sector (2015). *(Data from INECC, 2018. Investigaciones 2018–2013 en materia de mitigación del cambio climático. Retrieved April 16, 2018, from Instituto Nacional de Ecología y Cambio Climático: https://www.gob.mx/inecc/documentos/investigaciones-2018-2013-en-materia-de-mitigacion-del-cambio-climatico.)*

Waste sector emissions: 46 Mt CO$_2$e

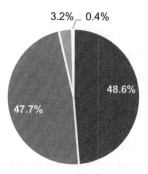

■ Wastewater treatment and disposal   ■ MSW disposal

■ MSW incineration and open burning   ■ Biological treatment of MSW

Fig. 2.11 GHG emissions from waste sector (2015). *(Data from INECC, 2018. Investigaciones 2018–2013 en materia de mitigación del cambio climático. Retrieved April 16, 2018, from Instituto Nacional de Ecología y Cambio Climático: https://www.gob.mx/inecc/documentos/investigaciones-2018-2013-en-materia-de-mitigacion-del-cambio-climatico.)*

Waste composition in Mexico, shown in Fig. 2.12, is characterized by a significant share of organic waste, as is the case in most developing countries.

As for waste collection coverage, in Mexico 93.4% of generated waste was collected in 2012, a figure that compares favorably to collection rates in developed countries.

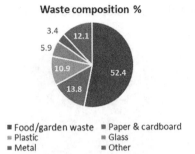

Fig. 2.12 MWS composition. *(Data from SEMARNAT, 2015a. Informe de la Situación del Medio Ambiente en México. Edición 2015. Capítulo.)*

Recycling rates vary according to the materials present in waste composition. In 2012, paper and cardboard reached a 32% recycling rate, while recycling of PET (polyethylene terephthalate) reached 15.8%, and glass 13.8% (although PET recyclers' association claims over 50% recycling rate for this material) (ECOCE, 2018). The national average rate for recycling was 9.6% in 2012 (SEMARNAT, 2016).

Finally, waste disposal facilities in 2013 in Mexico are 21% open dumps, while 74.5% goes to sanitary landfills (SEMARNAT, 2016).

2.1.3 Social Overview
2.1.3.1 Argentina—Social Overview

Alejandro Cittadino* and Atilio Armando Savino[†]

*Departamento de Ecología, Genética y Evolución—Facultad de Ciencias Exactas y Naturales, Universidad de Buenos Aires, Buenos Aires, Argentina
[†]International Solid Waste Association (ISWA), Buenos Aires, Argentina

According to the national census of 2010 in Argentina, the population was 40,117,096, which means an average density of $14.4 \, cap/km^2$, without considering Antarctica and the South Atlantic Islands (INDEC, 2010).

About 91.03% of the population is located in urban areas (more than 2000 inhabitants) and 47.4% of the population lives in urban agglomerates of more than 500.000 inhabitants, The Metropolitan Area of Buenos Aires (AMBA)[8] being the most important representing 31.9% of the total population. The rest corresponds to the Gran Córdoba (3.6%), Gran Rosario (3.1%), Gran Mendoza (2.3%), Gran San Miguel de Tucuman (2%), La Plata (1.6%), Mar del Plata (1.5%), and Gran Salta (1.34%) (INDEC, 2010). The Argentine provinces are also divided into departments and each department includes one or more municipalities, except the Province of Buenos Aires, which is divided into districts that match the number of municipalities. The CABA is organized into

[8] City of Buenos Aires and 24 districts of the Province of Buenos Aires.

communes. The number of districts, departments, and communes amounts to a total of 527 and the municipalities to 2165 (INDEC, 2010).

According to the HDI computed by the PNUD (2016), Argentina ranked 45th out of 188 countries. Historically, the poverty and indigence levels have been very high, with a minimum average of over 20% in the last 25 years (PNUD, 2017). In the second half of 2017, the percentage of households below the poverty line[9] was 17.9%, which comprises 25.7% of the population. Within this group, must be considered a 3.5% of indigent households,[10] that is, a 4.8% of the population (INDEC, 2018). Although Argentina is eminently an urban country, it presents a significant housing shortage, and alarming signs of residential segregation between the gated communities where high-income people live and slums and irregular settlements.

In terms of management, urban SW is responsibility of the municipalities. In general, there are no systematized statistical databases distinguishing the different stages of the waste management. In 2005 the Secretary of Environment and Sustainable Development, nowadays upgraded to Ministry, developed the "National Strategy for Urban Solid Waste" (ENGIRSU in Spanish) with the aim of reverting inadequate practices in the management of MSW, promoting the closure of open dumps, the waste reduction, recovery and recycling, and the construction of landfills. In this framework, several diagnostic screening studies and projects have been developed (ENGIRSU, 2005, 2016; ARS, 2012; Banco Mundial, 2015).

Regarding waste collection, 94.82% of urban households have this service; this percentage decreases to 89.91% when considering rural areas. The areas with the greatest deficiency correspond to the rural areas and within the urban areas, mainly slums and deprived urban neighborhoods that surround the city of Buenos Aires. Most of the provinces have a collection coverage greater than 80%, being the lowest rates in the Northeast region (provinces such as Santiago del Estero and Formosa have coverage percentages of 64.8% and 64.04%, respectively), followed by the provinces of the Northwest region (Banco Mundial, 2015; INDEC, 2010; ENGIRSU, 2016) (Fig. 2.13).

Taking into account the average coverage collection rate and considering the total population estimated in the last census and its 2018 projection, there would be more than 4 million people without service of regular collection.

The waste generation varies between 0.91 and 0.95 kg/capita/day with a maximum of 1.52 for the CABA and a minimum of 0.44 for the Province of Misiones (ENGIRSU, 2005). In 2012 the Solid Urban Waste Association (ARS, 2012) estimated an average waste generation of 1.022 kg/capita/day with a maximum and minimum in the CABA

[9] Percentage calculated considering the evolution of the basic food basket and the total basic basket, compared to the income of households.

[10] The concept "indigence line" establishes whether households have enough income to meet the food basket capable of satisfying the minimum nutritional requirements. In this way, households that not exceed that threshold or line are considered indigent.

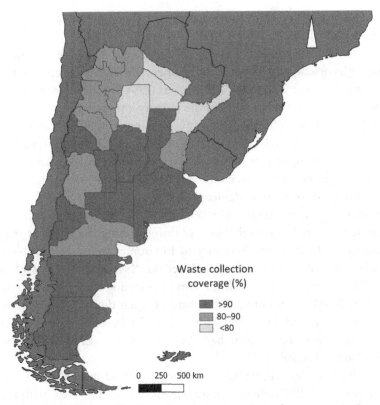

Fig. 2.13 Municipal solid waste collection coverage in percentages by province in Argentina. *(Data from Estrategia Nacional para la Gestión Integral de Residuos Sólidos Urbanos (ENGIRSU), 2016.)*

and Misiones (1.252 and 0.641 kg/capita/day, respectively). Finally, the Banco Mundial (2015) considers an average waste generation of 1.22 kg/capita/day. As previously mentioned, the lack of statistical records makes the estimation of waste generation per capita and its evolution over time difficult. However, the average of the different estimates previously cited indicates that the country's waste generation is about 1.03 kg/capita/day, being the provinces with higher income those that are above it, especially the CABA. It is important to note that, considering these information and population estimations from INDEC, Argentina would generate between 40,489,997 and 55,667,516 kg of MSW/day.

According to the survey carried out by the Banco Mundial (2015) there are 150 mechanized MSW separation plants in Argentina (Fig. 2.14), with a total operative capacity of 8665 ton/day. They work below their capacity, there are even plants that although were acquired never started to operate and others that closed due to lack of maintenance. Even without considering these cases, taking only into account the installed capacity, the maximum treated fraction would be 17.7% of the total waste

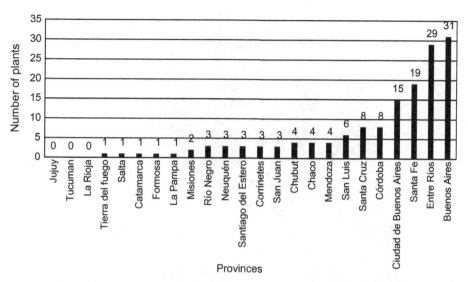

Fig. 2.14 Number of mechanized municipal solid waste separation plants by Province in Argentina. *(Data from Banco Mundial, 2015. Diagnóstico de la Gestión Integral de Residuos Sólidos Urbanos en la Argentina. Recopilación, generación y análisis de datos – Recolección, barrido, transferencia, tratamiento y disposición final de Residuos Sólidos Urbanos. The World Bank.)*

generated, about 49,070 ton/day (Banco Mundial, 2015). If the total number of separation plants, mechanized or not, is considered, then the total amount would be 187 (ENGIRSU, 2016).

Unfortunately, there are no statistics of the amount of waste treated, recovered, and finally marketed. According to Banco Mundial (2015), recovery rates for plants that separate inorganic waste are less than 10%. Rates are low due to the lack of separation programs at source, insufficient or ineffective awareness campaigns, operational problems, and lack of plant maintenance. There is also a considerable spatial heterogeneity with strong concentration of plants in the central region of the country, mainly in the provinces of Buenos Aires, Santa Fe, Entre Ríos, and Cordoba (see Fig. 2.14).

Finally, there is a mostly informal recovery system/circle organized by waste pickers, who salvages reusable or recyclable materials from the roadside and from the final disposition sites, like open dumps or landfills. Again, there is a lack of comprehensive statistical records. In the CABA, waste pickers are responsible for the recovery of 960 ton/day of the 6760 ton/day of waste generated (14.2%, according to official communications of the Government of the City of Buenos Aires).

The three largest metropolitan areas in the country, Gran Buenos Aires, Gran Córdoba, and Gran Rosario, have composting plants. There are no formal records of the total number of plants in Argentina. It is suspected that this activity is more widespread in small cities associated with waste separation plants (ENGIRSU, 2005).

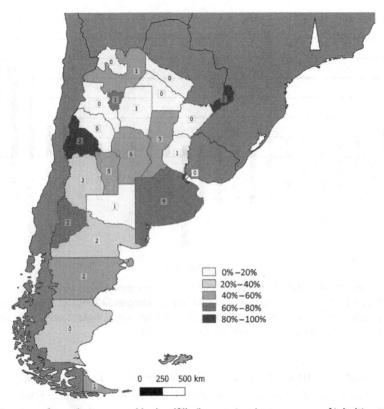

Fig. 2.15 Fraction of population served by landfills (by province): percentage of inhabitants that have an adequate final disposition service at provincial level, considering the total population of the municipalities that have a landfill over the total province population. Note: The numbers indicate the existing landfills. *(Source: ENGIRSU (2016).)*

According to official data (ENGIRSU, 2016), the population disposing MSW in landfills reaches 61% with great variability between the provinces (Fig. 2.15). There are six provinces that don't have any landfill sites (La Rioja, Jujuy, Catamarca, Formosa, Corrientes y Chaco), but the rest have at least one. The total number of operative landfills in the country is 42 (Banco Mundial, 2015) (Fig. 2.15).

The landfills of Argentina are located mainly in the main urban agglomerates and cities most visited by tourists. The greatest lack of coverage occurs in cities with low number of inhabitants: in municipalities with less than 15,000 inhabitants, only 9.4% of its population dispose adequately the waste generated, in cities where the inhabitants are between 15,000 and 50,000 the rate is 24.5% (Banco Mundial, 2015).

The City of Buenos Aires together with 40 municipalities in the Buenos Aires Metropolitan Region (including the 24 of the AMBA) that are part of the regional system under management of the state company CEAMSE (Coordinación Ecológica

Área Metropolitana Sociedad del Estado) represent 46% of the disposal in landfills in Argentina. This percentage rises to 80% when considering the central region of the country (rest of Buenos Aires Province, Córdoba, Santa Fe, Entre Ríos, and La Pampa), ENGIRSU (2016).

The municipal waste not disposed in landfills at best goes to controlled dumps, but in most cases end up in open dumps (ENGIRSU, 2016; Banco Mundial, 2015). On the other hand, even in metropolitan areas served by landfills, open dumps are common. For example, in the Metropolitan Area of Buenos Aires, there are 292 illegal open dumps larger than 1 ha (Cittadino et al., 2012).

2.1.3.2 Brazil—Social Overview

Vanessa Pecora Garcilasso and Fernando C. de Oliveira
Research Group on Bioenergy, Institute of Energy and Environment, University of São Paulo, São Paulo, Brazil

Sanitation is one of the biggest challenges in Brazil. The levels of collection, disposal, or adequate treatment of water, sewage, and garbage interact with other indicators, such as income and education, and point to consequences such as the persistence of high number of hospitalizations due to illnesses related to lack of basic sanitation, more frequently in the North and Northeast regions.

According to the latest version of the 2016 Panorama of the Solid Waste in Brazil, released in 2017 by the Brazilian Association of Companies for Public Cleaning and Special Waste, in 2016 Brazil generated about 78.3 million tons of solid urban waste, being collected only about 71.3 million tons, that is, 7 million tons were given an uncertain fate and inadequate disposal, being vectors of diseases and environmental pollution. Of all MSW collected in the country, only 58.4% are intended for landfills, and about 41.6% of MSW collected are sent to controlled landfills or dumps, where they do not receive the appropriate final treatment (ABRELPE, 2017).

Fig. 2.16 shows the relation between produced and collected residues in Brazil, per region, in 2016, while Fig. 2.17 shows the final disposal of the collected residues, by region.

Fig. 2.17 shows that only in the South and Southeast regions most of the collected wastes are intended for landfills. In other regions, most of the waste collected unfortunately still has its final destination inadequately disposed, via controlled landfills or dumps.

The coverage of MSW collection services increased from 90.8% in 2015 to 91.2% in 2016 of the generated volume (ABRELPE, 2017). However, the selective collection did not advance at the same rate, so the recycling rates were stagnant for a few years. As a result, there is an overload on the final disposal systems of these wastes.

It is noteworthy that since 2002 the garbage collector activity in Brazil has been recognized as a professional category, registered in the Brazilian Occupation Classification (CBO), as "Recyclable Material Collector." This new class of workers performs the function of collecting, transporting, sorting, pressing, storing, and trading these materials to be reused (MNCR, 2014). The problem today is not in legally acknowledging the

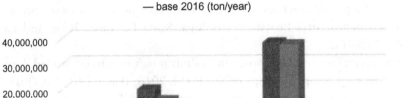

Fig. 2.16 Solid urban waste generated vs collected in Brazil, by region—base year 2016 (ton/year). *(Adapted from ABRELPE, 2017. Panorama dos Resíduos Sólidos no Brasil 2017.)*

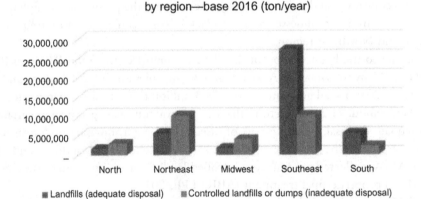

Fig. 2.17 Final destination of solid urban waste collected in Brazil, by region—base 2016 (ton/year). *(Adapted from ABRELPE, 2017. Panorama dos Resíduos Sólidos no Brasil 2017.)*

garbage collectors as professionals, but rather in granting their rights to decent conditions of work and life beyond the sole standpoint of survival.

Waste collectors are fundamental in the garbage collection process. Recycling in Brazil is done at the expense of the collector's work and not the selective collection. The fact that the garbage collectors are in the CBO could be an indicative of the redemption of these workers' dignity by placing them into the framework of public policies. Unfortunately, a contrasting situation is observed in the country. The collectors are exposed to health risks, social prejudices, and deregulation of labor rights, conditions that are extremely precarious, both in the informality of work and wage. In addition, they do not have access to formal education and technical improvements. To have an effective social inclusion, proposed by the Brazilian Policy of Solid Waste, not only the aspects of the right to work and income should be the focus of the actions of public authorities, but also the health conditions and the risks to which the recycling workers are exposed.

2.1.3.3 Ecuador—Social Overview

Rafael Soria and Laura Salgado

Departamento de Ingeniería Mecánica, Escuela Politécnica Nacional, Ladrón de Guevara, Quito, Ecuador

Ecuadorian population has increased from 12.4 million inhabitants in 2000 to 16.4 million inhabitants in 2016 (OLADE, 2017). Current annual average growth rate of population is estimated at 1.5%. Official population projections estimate 19.8 million inhabitants in 2030 and 23.4 million inhabitants in 2050 (INEC, 2012).

Historically, from 1982 to 2013 the average monthly family income was always lower than the value of the basic family consumption basket (INEC, 2017b). According to the national survey of income and expenses in urban and rural households (ENIGHUR) in 2013, only 58.8% of Ecuadorians had the capacity to save money, while 41.1% registered more expenses than income. The average of total monthly income in Ecuador was 893 USD/person in 2013, while the average monthly expenditure was 810 USD. In rural areas the situation is different; both statistics vary between 567 US$ and 526 USD, respectively. Only during the first semesters of 2014, 2015, and 2016 the opposite situation was verified. More recently, in December 2017, the basic familiar basket was valued at 708.98 USD, which is higher than the average monthly family income of 700 USD, in the same month (INEC, 2017b).

Although during the period 2007–17 the average of the relation between population with employment and population in economic activity age is 95.2%, the quality of this employment has decreased substantially from the second semester of 2012, when the relation between underemployed and employment was 9.4%, up to a level of 22.3% in March 2017 (INEC, 2017b). Recently the Ecuadorian economy has shown some signs of improvement, demonstrating an indicator of unemployment of 19.8% in December 2017 (INEC, 2017b). The multidimensional poverty index until December 2016 was 35.1%, while the monetary poverty until June 2017 was 23.1%.

In 2016, a total of 12,898 tons of MSW/day was collected at national level (AME/INEC, 2017), which represents an average rate of daily waste production per capita (PPC) of 0.73 kg/person/day (MAE, 2017b). In Ecuador 74.3% of the population live in regions with population concentrations of 2000 people or more, which is understood as urban regions (MIDUVI, 2016). The average PPC in urban regions is 0.58 kg/person/day (AME/INEC, 2017). In Quito, the capital of Ecuador, the PPC

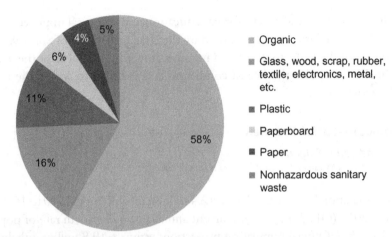

Fig. 2.18 Average composition of MSW in urban regions in Ecuador in 2016. *(Adapted from AME/INEC, 2017. Estadística de Información Ambiental Económica en Gobiernos Autónomos Descentralizados Municipales – Gestión de Residuos Sólidos 2016. INEC, Quito, Ecuador. Available from: http://www. ecuadorencifras.gob.ec/documentos/web-inec/Encuestas_Ambientales/Gestion_Integral_de_Residuos_ Solidos/2016/Documento%20tecnico%20Residuos%20solidos%202016%20F.pdf.)*

is 0.74 kg/person/day (EMASEO, 2014). The average composition of MSW in urban regions in 2016 is shown in Fig. 2.18. The average humidity content of urban MSW in Ecuador is 51% and the average low heat value (LHV) is 5.5 MJ/kg (TNA/MAE/URC/GEF, 2013).

Ecuador has 221 Decentralized Autonomous Governments (GADs) of which 38% of GADs use landfills for the final disposition of MSW, while the remaining GADs use garbage dumps (50%) and provisional cell (12%) (see Fig. 2.19). According to the National Population Census 2010, 62.7% of households deliver the residues to the garbage collector truck, 14.7% throw away the garbage to vacant lands, 17.9% incinerate garbage, and the remaining households use other forms of final waste disposition (INEC, 2010).

One of the main threats that the GADs face for the integrated management of MSW is its economic management model. Currently in most of the GADs the rate charged by the MSW recollection system and final disposal is lower than the cost of the service, thus there is an important subsidy financed by the GADs. In this sense, 48% of GADs collect this rate through the electricity bill, 24% through the drinking water service and sewage water bill, 10.4% do not have regulations to collect this rate, 14% through direct payments to the municipality, and 3.6% through the property tax (AME/INEC, 2017).

The incorporation of elements of differentiated collection, recycling, treatment, and use of residues in the operations of GADs is a priority for the Ministry of Environment (MAE), led by the National Program of Integrated Management of Solid Residues

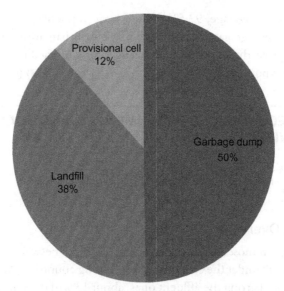

Fig. 2.19 Final disposition of MSW in Ecuador by GADs. *(Adapted from AME/INEC, 2017. Estadística de Información Ambiental Económica en Gobiernos Autónomos Descentralizados Municipales – Gestión de Residuos Sólidos 2016. INEC, Quito, Ecuador. Available from: http://www.ecuadorencifras.gob.ec/documentos/web-inec/Encuestas_Ambientales/Gestion_Integral_de_Residuos_Solidos/2016/Documento%20tecnico%20Residuos%20solidos%202016%20F.pdf.)*

(PNGIRS). Some of the GADs that succeed in carrying on vermiculture and composting are: Cuenca, Loja, Macas, Mejía, Atuntaqui, Otavalo, and the commonwealths of Patate-Pelileo and *Mundo Verde*.

2.1.3.4 Mexico—Social Overview

Gustavo Solórzano
Asociación Mexicana de Ingeniería, Ciencia y Gestión Ambiental, A.C. (AMICA), Mexico City, Mexico

In 2017, the total population in Mexico was estimated at 123.5 million inhabitants (CONAPO, 2017). Available figures for 2015 indicate that 23% of the population was considered rural and 77% lived in urban areas, while 39.3% of the total population (or 47.8 million people) concentrated in 14 metropolitan zones with over 1 million inhabitants each. Around 53% of the population lives in locations above 1500 m (CONAGUA, 2016), although most hydraulic resources in the country are only available below this altitude.

The HDI for Mexico has been reported as 0.756 for 2014; life expectancy at birth for same year was 76.8 years (UN, 2018). An estimated 55.3 million people or 46.2% of the population in Mexico lived in some degree of poverty in 2014. Of this figure, an estimated 11.4 million people lived in conditions of extreme poverty (CONAGUA, 2016).

As for water supply coverage, 92.5% of the Mexican population is served with piped potable water (with disaggregated figures of 95.7% for urban areas, and 81.6% for rural areas). The coverage for the population accessing sanitation facilities (sewerage systems) reached 91.4% of the population (96.6% urban, 74.2% rural areas) (CONAGUA, 2016).

2.2 ASIA: ECONOMIC, ENVIRONMENTAL, AND SOCIAL OVERVIEW

Shyamala K. Mani* and Bini Samal[†]
*National Institute of Urban Affairs (NIUA), India Habitat Centre, New Delhi, India
[†]Forest Research Institute, Dehradun, India

2.2.1 Economic Overview

It is well known that in most Asian countries, the GDP lies between 2.5% and 4.5% and hence most of them fall under the category of developing countries (ADB, 2017). In most of the Asian countries, barring the affluent ones, about 13% of the population lives in low income and densely populated areas where sanitation and waste management services are not as per their country's guidelines.

The consequences of burgeoning population in urban centers are more noticeable in developing countries as compared to the developed countries. In India, about 31.2% population is now living in urban areas. Over 377 million urban people are living in 7935 towns/cities (Indian Census, 2011). The population residing in urban regions increased from 18% to 31.2% from 1961 to 2011, respectively (Indian Census, 2011).

India is a vast country divided into 29 states and 7 union territories (UTs). There are three mega cities—Greater Mumbai, Delhi, and Kolkata—having a population of more than 10 million; 53 cities have a population over 1 million, and 415 cities have a population of 100,000 or more, but less than 1 million (Indian Census, 2011). It is projected that in Asia, India will continue to have a high population growth rate, surpassing Chinese population in the middle of the 2020s. India will be world's most populous country at about 1.6 billion in 2040. Countries in the Association of Southeast Asian Nations (ASEAN) have population larger than the European Union and its increase will be only second to India's.

China's population, currently the world's largest, around 2030, will be 1.41 billion but will decrease to 1.21 billion by 2040. China has more than 100 million elderly people aged 65 or more and this number will increase. In China, rural population aging is perceived to be a serious problem since the younger population would be in urban regions. There would be a continuous population increase in Asia as a whole. The share of Asia's population globally would however fall from 55% in 2014 to 50% in 2040.

After the United States and Europe, the Chinese economy, although currently the third largest in the world, has reduced its growth rate to about 7% due to reduced investment, thus slowing exports. This is because of economic slowing down in Europe and resource-rich countries. Asian emerging economies too have slowed down due to Chinese economic deceleration, which had earlier expanded because of rising exports to China. Singapore, Malaysia, and Thailand that focused on large share of their respective exports to China have also now slowed down because of the decline in Chinese demand. Since the world economy and the US economy are improving, emerging economies in Asia will improve. Downward pressure is because of weak oil prices exerted on Russia as well as Middle Eastern and Latin American countries, which are oil producing as well as resource-rich countries.

Many economies including those in Asia will rebound through concerted medium and long-term international actions using appropriate fiscal and monetary policies. India is projected to be the new driver of global economic growth and will make its presence felt in the global arena. The Indian economy is projected to grow at an annual pace of 6.2%, fastest in the world. Foreign direct investment, domestic demand expansion, and structural reform would become India's sources of economic growth. India's exports to China accounted for only 0.7% of its gross domestic product and 4% of its total exports. Therefore, India won't be affected by China's slowing down of economic growth. China too will retain an annual economic growth rate of 5.1%. The annual growth rate of ASEAN economy will be 4.5%.

There is no doubt that the current environment has changed because of rising wages and citizens' growing consciousness of rights, from abundant surplus labor and low costs which drove export-oriented growth in Asia in the previous century to demand-driven economies currently. However, Asia will remain the center of global economic growth. Although there is no indication of limit on Asia's economic growth, countries in Asia must take precautions against the so-called *middle-income country trap*.

2.2.2 Environmental Overview

Agamuthu Pariatamby*, Bini Samal[†], and Shyamala K. Mani[‡]
* University of Malaya, Kuala Lumpur, Malaysia
[†] Forest Research Institute, Dehradun, India
[‡] National Institute of Urban Affairs (NIUA), India Habitat Centre, New Delhi, India

As urbanization and economic development increase in Asia, nowhere is the impact more obvious than in society's SW. The urban areas of Asia produce about 760,000 tons of MSW per day, or approximately 2.7 million m^3 per day. In 2025, this figure will increase to 1.8 million tons of waste per day, or 5.2 million m^3 per day. These estimates are conservative; the real values are probably more than double this amount (What a waste: Solid Waste Management in Asia, World Bank, 1999). Gross domestic product (GDP) in Asia has expanded by 5.7% in 2016 and 2017. Inhabited by more than 4.45 billion people in 2016, equivalent to 59.78% of total world population, Asia recorded a huge amount of waste generation, making it the largest waste-producing continent on Earth. By 2025, it is estimated that 1.8 billion ton will be generated by urban cities alone in Asia (Fig. 2.20).

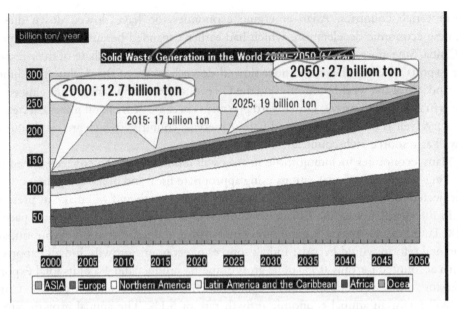

Fig. 2.20 Solid waste generation in the World until 2050. *(Source: Prof. P. Agamuthu, University of Malaya (Personal communication) (2018).)*

High-income countries produce the most waste per capita, while low-income countries produce the least SW per capita as illustrated in Fig. 2.20. Waste composition is influenced by factors such as culture, economic development, climate, and energy sources. Low-income countries have the highest proportion of organic waste. Paper, plastics, and other inorganic materials make up the highest proportion of MSW in high-income countries. Agricultural wastes are also very common in Asian cities and their quantities are shown in Fig. 2.21.

Although waste composition is usually provided by weight, as a country's affluence increases, waste volumes tend to be more important, especially regarding collection: organics and inert generally decrease in relative terms, while increasing paper and plastic increases overall waste volumes. This is depicted in Figs. 2.22–2.24 . It is also important to note that a lot of food gets wasted while moving across different stages of food supply chain. This is shown in Fig. 2.25.

Primary waste collection services in Asian countries are often managed by community-based organizations or small enterprises and require the residents to pay monthly collection fees. The municipality would handle the intermediate collection, siting transfer, or even disposal. However, with rapidly increasing waste and high cost of transport or disposal, municipality usually encourages households to recycle as much as possible so that there is very little need for transport of collected waste (Zurbrugg, 2002).

For collection and transportation of SW in Asian developing countries, SW cycles through collection, transportation, and final disposal. In Jakarta only 70% waste was

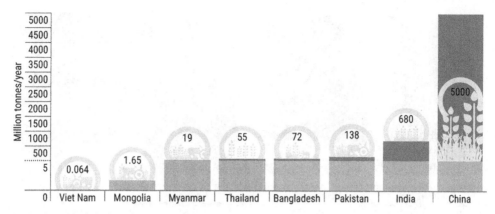

Fig. 2.21 Agricultural waste generated by selected Asian countries.

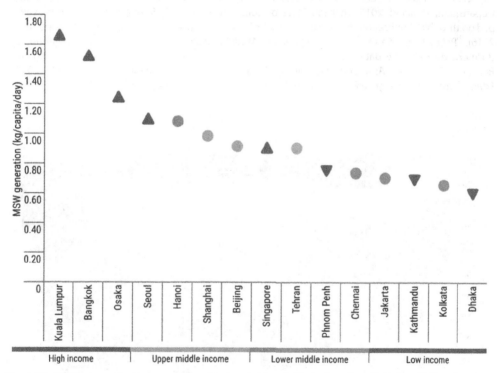

Fig. 2.22 MSW generation per capita by city income level. *(Data from Kawai, K., Tasaki, T., 2016. Revisiting estimates of municipal solid waste generation per capita and their reliability. J. Mater. Cycles Waste Manag. 18, 1–13. https://doi.org/10.1007/s10163-015-0355-1.)*

collected in 2007 (Pasang et al., 2007) while collection is improving currently. Despite this, mechanical equipment is used infrequently, and manual collection is more common for picking up the SW (Moghadam et al., 2009). The problem of SW collection is caused because of the nonavailability of transfer station facility as in Tibet (Jiang et al., 2009).

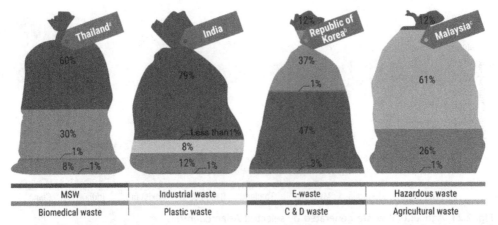

Fig. 2.23 MSW as a fraction of total wastes generated in select Asian countries. [a]Pollution Control Department, Thailand, 2015. Thailand State of pollution report—2015. Available from http://www.pcd.go.th/public/Publications/en_print_report.cfm?task=en_report2558. (Accessed 30 September 2016). [b]Policy Direction of Resource Circulation. Available from http://eng.me.go.kr/eng/web/index.do?menuId=364. [c]The data was extracted from the desktop research file conducted by Regional Resource Centre for Asia and the Pacific, Asian Institute of Technology (AIT RRC.AP, 2017). *From Chapter 2: Waste generation. In: Asia Waste Management Outlook, p. 27.*

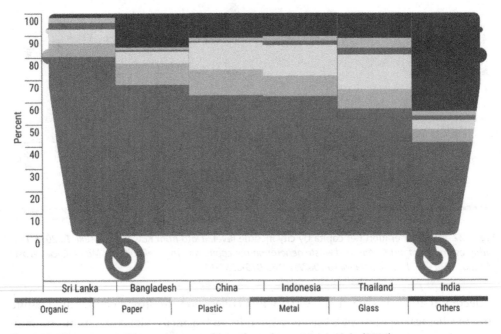

Fig. 2.24 Waste composition as generated by selected countries in Asia (2016).

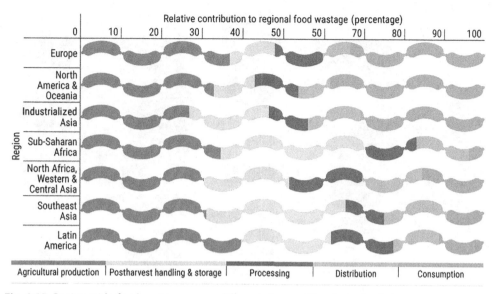

Fig. 2.25 Region-wide food wastage across different stages of food supply chain.

The collection service is now door-to-door, such as in Jakarta (Pasang et al., 2007), in metro cities in India (Kumar et al., 2009) especially in Bangalore, etc. (Ramachandra and Bachamanda, 2007).

As per a report by Hoornweg and Bhada-Tata (2012a,b) it has been observed that low-income countries continue to spend most of their MSW budgets on waste collection, with only a fraction going toward disposal. This is opposite in high-income countries where the main expenditure is on disposal. Fig. 2.26 shows the common MSW disposal methods in Asia by country income level.

As stated by UNEP (2004), composting is one of the treatments for SW, which is more suitable than other treatments in Asian developing countries especially incinerators (Meidiana and Gamse, 2010). The component of MSW that is in abundance in those countries is decomposable organic matter, which has high moisture content. The constraints of composting in Asian developing countries includes high cost in operation and maintenance and poor maintenance and operation of facilities, incomplete separation of noncompostable materials, etc. Besides, higher cost of compost compared to commercial fertilizers also affects the implementation of composting (Tchobanoglous et al., 1993). The MSW in Asian developing countries is also plagued with low financial investment and low enforcement of environmental regulation (Visvanathan and Glawe, 2006). As to achieving targets of composting, in India composting is about 22% (MoHUA), and in other countries like Nepal, Pakistan, Bangladesh, and Sri Lanka it is less than 10% (Khajuria et al., 2010).

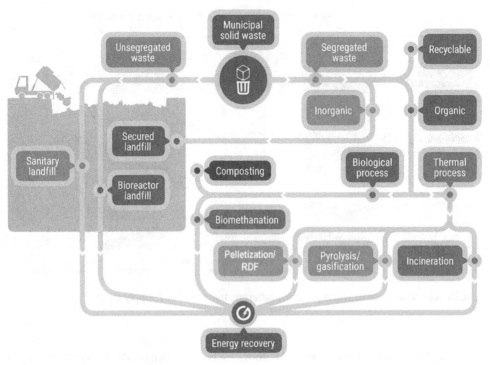

Fig. 2.26 Common municipal solid waste disposal methods in Asia by country income level.

Table 2.4 Municipal Waste Facilities in Selected Asian Countries

Country	Treatment Plants	Incinerators	MRF	Open Dumpsites	Controlled Landfills	Sanitary Landfills
PR China	419	69	NA	NA	324	20
Indonesia	20	0	80	400	70	10
Republic of Korea	4955	2028	0	325	1348 (which includes solidification and gasification)	
Malaysia	NA	4	1	261	10	12
Philippines	NA	26	2361	826	273	19
Singapore	NA	4	1	–	–	1
Thailand	NA	3	NA	NA	20	91
Viet Nam	NA	NA	NA	49	91	17

Source: Regional Resource Centre for Asia and the Pacific (2010).

Methods for final MSW treatment and disposal in developing Southeast Asian countries are commonly open dumping, landfill, and others (Table 2.4). There were proportions for various processes, namely, open dumping (more than 50%), landfill (10%–30%), incineration (2%–5%), and composting (<15%). The final disposal method is generally

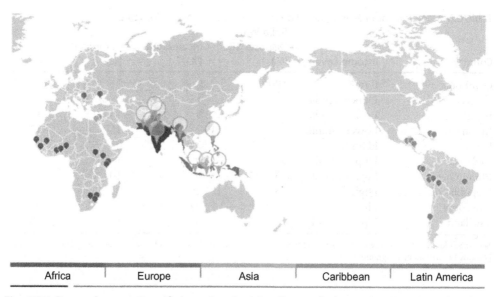

| Africa | Europe | Asia | Caribbean | Latin America |

Fig. 2.27 Size and proportion of dumpsites in Asia. *(Source: P. Agamuthu, University of Malaya, Malaysia.)*

open dumped landfill. In Malaysia the amount of SW collected for final disposal was about 70%, whereas 20%–30% was dumped or thrown into rivers (Meidiana and Gamse, 2010). Almost similar conditions were found in Indonesia, where dumped landfill was the primary method of disposal. In Indonesia, the SW transported to dumped landfill was 69%, buried 9.6%, composted 7.15%, burnt 4.8%, disposed off in the river 2.9%, and others 6.55% (Meidiana and Gamse, 2010). Out of the largest 50 dumpsites in the world, 17 dumpsites are found in Asia as shown in Fig. 2.27.

Some alternative solutions that have been successfully implemented in Surabaya and Medan, Indonesia are as follows. In these cities, public awareness was improved after receiving guidance concerning environmental issues. The trainers were local leaders and facilitators with the assistance of nongovernmental organizations (NGOs). This program was performed as community-based MSW (USAID, 2006). The program successfully applied 3Rs (reduce, reuse, and recycling), which included waste separation at the source and composting. A Refuse Bank, which received domestic recyclable waste from the community, has also been initiated and is operated in these cities. In Yala, Thailand, the poor communities reducing SW was triggered by exchanging of the trash for nutritional food (Mongkolnchaiarunya, 2005).

Composting is one of the preferred methods for reducing biodegradable organic material. Composting can reduce more than 50% of biodegradable organic components of SW on-site. Composting decreased the residential MSW between 38% and 55% in Dar es Salaam City, Tanzania (Mbuligwe et al., 2002).

Table 2.5 Management of Municipal Solid Waste in Selected Asian Countries

Country	Income Level	Solid Waste Disposal Site (%)	Incineration (%)	Composting (%)	Other (%)
Cambodia	Low	100	0	0	0
China	Upper middle	85	15	0	0
India	Lower middle	75	5	10	10
Indonesia	Lower middle	70	2	15	13
Japan	High	3	74	0	17
Malaysia	Upper middle	93	0	1	6
Philippines	Lower middle	85	0	10	5
Republic of Korea	High	35	28	37	0
Singapore	High	6	94	0	0
Thailand	Upper middle	70	5	10	15

Source: Seadon, J., 2017. Waste economics and financing. In: Modak, P. (Eds.), Asia Waste Management Outlook. United Nations Environment Programme.

Some developing countries in Asia are trying to change MSW into energy. India, Philippines, and Thailand have converted waste to energy. In Yala, Thailand, a program of recycling and garbage reducing was established through a relationship between poor communities and the municipal administration (Mongkolnchaiarunya, 2005). Similarly, in India, the government had established cooperation with private sector and citizens in recycling SW (Zurbrugg, 2002). On the other hand, initiatives of the private sector (citizen and enterprises) such as public-private-community partnership also helps to increase the efficiency of waste management systems (Zurbrugg, et al., 2004) (Table 2.5).

2.2.2.1 Challenges of MSW Management in Asian Developing Countries

The main challenges on MSW management in Asian developing countries are related to appropriate waste disposal, and some figures are significant:

- Only 10%–15% of the disposal sites are sanitary landfills.
- In Indonesia and Vietnam, they are mostly nonengineered landfills.
- There is a lack of appropriate technology to be applied, such as lack of lining system, landfill gas collection system, which makes these dumpsites hazardous.
- Most dumpsites are a source of environmental pollution.
- Leachate, landfill gas, pest, vermin, and scavengers abound at most dumpsites.

The main problem of landfill managers is that there are no related policies in many Asian developing countries regarding what and how much can be disposed into landfill sites. Lack of knowledge of what materials are being dumped into landfills leads to the generation of leachates and landfill gas, which if not properly estimated can lead to a dangerous situation which can explode at any time.

In many developing Asian countries, huge hillocks of garbage on the outskirts of cities can be seen smoldering continuously from the emission of methane and other gases generated by the burning of plastics and nonbiodegradable materials in the dump. This contributes substantially (up to 11%–12%) to particulate matter pollution as well as gaseous emissions harmful to health. Lack of segregation at source and processing leads to illegal dumping of waste on roadsides, riversides, and forestlands leading to soil and water contamination.

Furthermore, illegal dumping of waste also attracts pests and vermin, which become transmitters of infectious diseases, a major public health issue in developing countries.

2.2.2.2 Pest and Vermin

- Putrescible waste attracts insects and other animals.
- Common insects such as flies and mosquitoes breed in water that gets accumulated in discarded containers, which are part of waste.

Common animals such as domestic animals (dogs, cow, goat, etc.), monkeys, birds, etc., are attracted to easy and abundant food sources in the waste, which can lead to disease outbreaks, which then become difficult to control.

2.2.3 Social Overview

Shyamala K. Mani* and Bini Samal[†]

* National Institute of Urban Affairs (NIUA), India Habitat Centre, New Delhi, India
[†] Forest Research Institute, Dehradun, India

Waste workers and pickers in developing countries are seldom protected from direct contact and injury and the co-disposal of hazardous and medical wastes with municipal wastes poses serious health threat (Alam and Ahmade, 2013). The US Public Health Service identified 22 human diseases that are linked to improper solid waste management (Singh, 2013). To address these issues in such areas, master planning are necessary to improve services, especially sanitation, and waste management.

Current global MSW generation levels are approximately 1.3 billion ton per year and is expected to increase to approximately 2.2 billion ton per year by 2025 (Srivastava et al., 2014). This represents a significant increase in per capita waste generation rates, from 1.2 to 1.42 kg per person per day in the next 15 years (Srivastava et al., 2014). As urbanization and economic development increase in Asia, nowhere is the impact more obvious than in society's SW.

According to the University of Malaya, global MSW generation exceeded 17 billion ton in 2015 and is expected to reach 27 billion ton in 2050. Furthermore, besides the average generation going up to 1.42 kg/capita/day (ranging at 0.1–0.8 ton/capita/year), which mandates urgent need for sustainable waste management, adaptation of waste management hierarchy of 3Rs, workable and effective in many developed nations,

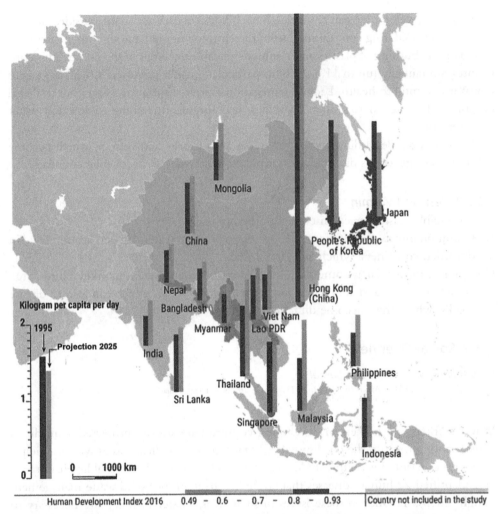

Fig. 2.28 MSW generation and projection for selected countries in Asia as linked to human development index. *(Source: UNDP, 2016. Human Development Report 2016 Human Development for Everyone. United Nations Development Programme.)*

was unsuccessful in many developing countries. The drivers of 3Rs practices need to be studied and enhanced in order to promote 3Rs in the developing nations. Fig. 2.28 depicts the MSW generation and its projected values in comparison to its HDI for various countries in Asia.

Although the 3Rs are considered the best ways to reduce waste generation and enhance its reuse and recovery, there are many challenges in achieving them. Some of the issues are listed below.

Issues of 3Rs in developing nations

1. 3R habit is achieved through stringent policy and regulation in Singapore, Japan, and Korea. This may not be happening in other Asian countries.
2. Waste is a livelihood for the urban poor communities in India, Bangladesh, and Indonesia. Hence stringent policy regulation implementation is a challenge.
3. Plastic recycling in India and Bangladesh reaches approximately 47% and 51%, respectively. Hence regulations for curtailing use of plastics are often not successful. Citizens feel that since plastics are being collected and recycled, there is no need to reduce consumption.
4. With the improvement in the standard of living and availability of other livelihood options, recycling of plastics may come down leading to accumulation of plastics.
5. Belief that recycling is not worth practicing will make recycling rate reduce. This is being observed in many rapidly developing countries like Malaysia and Thailand.

Other factors that influence 3R practice

- Impracticality of recycling due to the absence of waste separation
- Lack of a clear policy and necessary enforcement
- Nonsupportive local facilities
- Issues of trans–boundary movement of waste
- Absence of public participation
- Low levels of awareness on the benefits of practicing the 3Rs
- Consideration of informal recycling activities (scavengers and waste pickers) (Fig. 2.29)

Fig. 2.29 Ragpickers in Payatas landfill in the Philippines. *(Source: P. Agamuthu, University of Malaya, Malaysia.)*

Fig. 2.30 Drivers of 3R. *(Source: P. Agamuthu, University of Malaya, Malaysia.)*

Drivers toward the success of 3Rs (Fig. 2.30)
Current resource recycling strategies in Asia
1. 3R strategy has been so successful in Korea and Singapore that it has reduced MSW generation by approximately 22% and 10%, respectively.
2. Implementation of effective national waste management policies.
3. Other developing countries in Asia and the Pacific Islands reported unsuccessful story and instead rapid increase in waste generation is seen.
4. Commingled waste is another major problem.

2.2.3.1 Waste Management in India

Waste is a by-product of living and is being generated at a faster rate than urbanization in India (Srivastava et al., 2014). Planning Commission Report (2014) reveals that 377 million people residing in urban area generate 62 million tons of MSW per annum currently and it is projected that by 2031 these urban centers will generate 165 million tons of waste annually. It is also projected that by 2050 it could reach 436 million tons. To accommodate the amount of waste generated by 2031, about 23.5×107 cubic meter of landfill space would be required and in terms of area, it would be 1175 ha of land per year. The area required from 2031 to 2050 would be 43,000 ha for landfills piled to a height of 20 m. These projections are based on 0.45 kg/capita/day waste generation (Planning Commission Report, 2014).

The SW management has been hitherto given a low priority in most of the developing countries and hence funds allocated by the government agencies for managing the waste are inadequate, resulting in poor environmental conditions (health and hygiene). Furthermore, the level of tax and tax collection is poor coupled with the unwillingness of citizens to pay for services, thus economically constraining efficiency of this sector. Hence, it is important to rationalize the expenditure and management of the resources sustainably at the local body level for SW management to become effective in a country like India.

The MSW is a state subject included in the 12th Schedule of the Constitution (74th Amendment) Act of 1992 and ULBs (urban local bodies) are mandated to provide MSWM in all urban areas. State laws governing the ULBs also stipulate MSWM as an obligatory function of the municipal governments. Despite 15 years of implementation of the Municipal Solid Waste Management Rules 2000, ULBs were not able to put in place good systems. Sustainable solid waste management (SSWM) is a people management issue and overemphasis of technological solutions to solving the MSW problem will only delay in realizing good results (Mani and Singh, 2016). The Solid Waste Management Rules 2016, which have been in force since April 2016, place greater emphasis on citizens' participation and use of decentralized technologies and management practices.

Waste Management in Delhi

Delhi is the most densely populated and urbanized city of India. The annual growth rate of population during the last decade (1991–2001) was 3.85%, it almost double the national average. Currently the inhabitants of Delhi generate about 7000–10,000 ton/day (TPD) of MSW, which is projected to rise to 17,000–25,000 ton/day by 2021 (Talyan et al., 2008). The MSW management has remained one of the most neglected areas of the municipal system in Delhi. About 70%–80% of the generated MSW is collected and the rest remains unattended on streets or in small open dumps. Only 23.2% of the collected MSW is treated through composting, WtE etc., and rest is disposed in uncontrolled open landfills at the outskirts of the city (MoHUA, 2017).

The major issues of concern are as follows (Hoornweg and Bhada-Tata, 2012a,b):
- Increased waste generation of about 2%–3% per annum (from a 2012 baseline).
- Complex waste composition including 1%–2% hazardous materials.
- Ineffective mechanisms to tackle this problem.
- Lack of public participation.
- More importantly, lack of proper policy framework in many countries. Most developing nations in the world still dispose of waste in landfill or dumpsites.

Scavenging by informal sector
- Waste is a source of livelihood for the low-income group
- Additional side income for municipal workers
- They retrieve valuable/recyclable materials
- However, they face many health hazards
- Also reported to start fire to collect metal-based wastes

Conclusions for Asian countries

The current 3R practices in Asia and the Pacific Islands differ from one country to the other.

- Successful stories are in economically developed nations such as Korea, Japan, and Singapore, while 3Rs are almost insignificant in other developing nations in Asia.
- Positive drivers of 3R implementations include appropriate human attitude and the economic drivers, strengthened with suitable directive and legislation.
- Negative factors on the other hand are the lack of human attentiveness, discouraging economic scenario, and absence of appropriate regulations pertaining to 3R practice.
- Thus, improvement to amend these negative factors is very crucial in order to ensure that implementation of 3Rs in Asia and Pacific Island can be sustainable.

2.3 AFRICA: ECONOMIC, ENVIRONMENTAL, AND SOCIAL OVERVIEW

William H.L. Stafford

Council for Scientific and Industrial Research, Stellenbosch, South Africa; Department of Industrial Engineering, University of Stellenbosch, Stellenbosch, South Africa

2.3.1 Economic Overview

African countries have long faced challenges related to economic growth that is required to reduce poverty and GDP per capita is among the lowest in the world (Fig. 2.31).

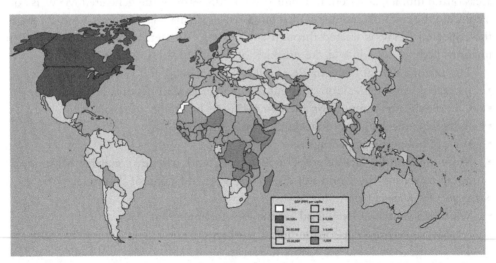

Fig. 2.31 GDP per capita. *(Source: https://commons.wikimedia.org/wiki/File:Gdp_per_capita_ppp_world_map.PNG.)*

For example, the average economic growth between 1990 and 2000 was only 2.1% per year. This is less than the population growth of 2.8% per year, and substantially less than the estimated 7% growth needed to reduce the proportion of Africans living in poverty by half by 2015.

There has been substantial progress in terms of macroeconomic stabilization but institutional weaknesses and poor governance as well as the reliance on primary production and primary products have undermined the development of secondary tertiary economies and stifled overall economic development. For example, approximately 60% of all exports from Africa are agricultural sector but they account for only 8% of the GDP. Fig. 2.32 shows GDP growth in selected African countries.

Historically, poor economic performance in the 1970s and 1980s led to the formulation of structural adjustment programs in many African countries that are composed of trade and payment liberalization, incentives for extraction of natural resources, privatization, and the removal of subsidies on social services such as education, health, and utilities. The poor governance characterized by a lack of transparency, accountability, and corruption increased poverty and led to human and financial capital being drained from Africa (Tutu, 1993).

Since the slump in the 1990s, Africa's economic growth has generally begun to improve with real rates of growth between 1% and 3%. However, performance in many countries was not encouraging. In addition, despite an increase in economic performance, per capita incomes have continued to remain stagnant that has resulted in increasing levels of poverty. However, it is encouraging that most African countries have achieved positive real growth rate for their economies since 2000. Africa achieved an average real GDP growth rate of 5.2% in 2004, 5.3% in 2005, and 5.7% in 2006. These improvements were generally underpinned by macroeconomic growth due to strong global demand for key African export commodities and are not a result of an increase in foreign direct investment that has been in decline since the 1970s. The foreign direct investment as a percentage of GDP for Northwestern east African countries has averaged around 1% per year, with only Central and Southern Africa obtaining greater foreign investment. This indicates a significant opportunity to improve the business environment and economic development in Africa through investment.

2.3.2 Environmental Overview

As economies develop and industrialize they historically made increasing use of energy including use of fossil fuels. In addition, development has typically come at a cost of natural resources—as can be seen by the ecological deficit attributed of developed nations (see Fig. 2.33).

Consequently, carbon emissions are typically reflective of industrial development status and this trend is observed for African countries. Only South Africa and Libya have per

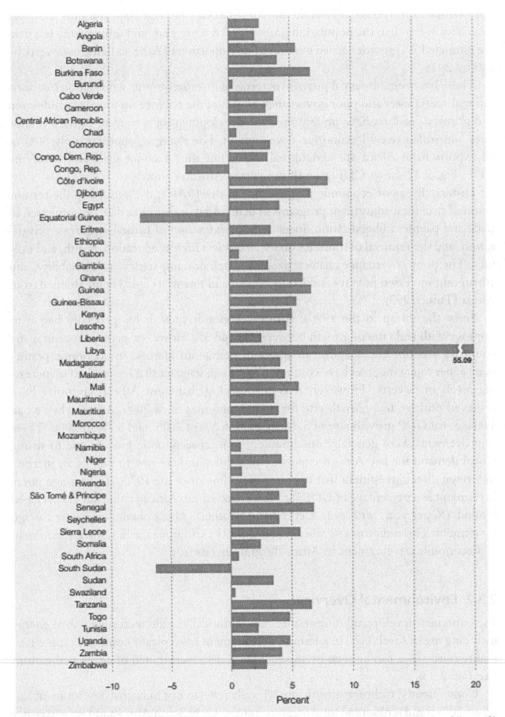

Fig. 2.32 GDP growth in selected African countries. *(Source: AfdB statistics; https://www.afdb.org/fil eadmin/uploads/afdb/Documents/Publications/African_Economic_Outlook_2018_-_EN.pdf.)*

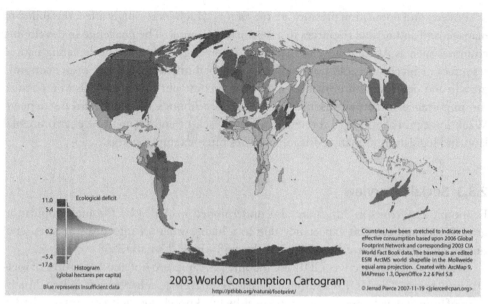

Fig. 2.33 World consumption cartogram showing ecological deficit. *(Source: http://pthbb.org/natural/footprint/2003/cartogram.png.)*

capita carbon dioxide emissions above the world average; moreover, in general, those countries with lower per capita GDP often tend to have lower per capita emissions. Aside from the GHG emissions from the use of fossil fuels to power economic development, significant GHG emissions are generated because of the clearing of forests and natural vegetation to convert land to agriculture and grazing. Additional sources of GHG emissions include tillage, savanna burning, use of fertilizer, and the disposal of organic wastes and sewage. For example, the highest methane emissions in Africa between 1990 and 1995 was attributed to Ethiopia which has one of the highest before headcount in the subregion and also notable methane emissions from the municipal dump in Addis Ababa that was estimated to be 9 Gg in 1998 (UNEP, 2015).

Although Africa contributes relatively little to global GHG emissions the region is extremely vulnerable to the impact of projected climate change since the majority of the population is highly dependent on natural resources and agriculture and poverty significantly limits the ability to respond and adapt to climate change. Also of increasing concern is the degradation of land in South Africa that is estimated to be about 500 million ha since 1950. This land degradation includes the depletion of nutrients erosion and damage to soil structure because of tillage, overgrazing, increasing application of chemicals, the use of inappropriate equipment and technologies, and commercial monocultures coupled with inefficient irrigation systems.

Poverty and population pressure are the factors which increasingly affect the ability to manage land and natural resources in a sustainable manner. The challenge in developing countries such as Africa is how to increase economic growth without depleting natural resources or increasing pollution and environmental impacts. In many cases economic growth and development prioritizes increasing income per capita and fails to recognize the importance of natural resources on which livelihoods of many Africans depend. Weak institutional and legal frameworks or the lack of enforcement of environmental laws and legislation in many African countries often exemplifies this.

2.3.3 Social Overview

Economic challenges in Africa are also underpinned in Africa by the high population growth rate and low life expectancy due to a high burden of infectious diseases, civil unrest, and poor health-care services, as shown in Figs. 2.34 and 2.35.

From a social perspective, GDP and income per capita are not good measures of societal well-being. The HDI is a summary measure of average achievement in key dimensions of human development: a long and healthy life, knowledge, and a decent standard of living. The HDI is a composite statistic (composite index) of life expectancy, education, and per capita income indicators, which are used to rank countries into four tiers of human development. A country scores higher HDI when the lifespan is higher, the education level is higher, and the GDP per capita is higher. Most countries in Africa have a low HDI (see Fig. 2.36).

Population growth, urbanization, a growing middle class, and changing consumption patterns drive SW generation in Africa. Changes in living standards and increases in

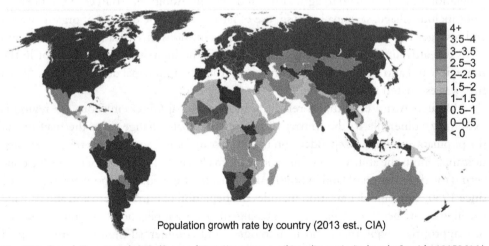

Population growth rate by country (2013 est., CIA)

Fig. 2.34 Population growth (%). *(Source: https://commons.wikimedia.org/w/index.php?curid=18159616.)*

Fig. 2.35 Life expectancy (years). (*Source: https://commons.wikimedia.org/w/index.php?curid=18159616.*)

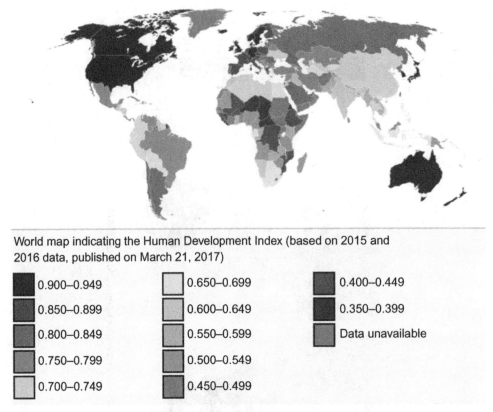

World map indicating the Human Development Index (based on 2015 and 2016 data, published on March 21, 2017)

■ 0.900–0.949	□ 0.650–0.699	■ 0.400–0.449
■ 0.850–0.899	□ 0.600–0.649	■ 0.350–0.399
■ 0.800–0.849	□ 0.550–0.599	□ Data unavailable
■ 0.750–0.799	■ 0.500–0.549	
□ 0.700–0.749	■ 0.450–0.499	

Fig. 2.36 Human development index. *(Sources: https://commons.wikimedia.org/wiki/User_talk:Happenstance; http://hdr.undp.org/sites/default/files/2016_human_development_report.pdf.)*

disposable income result in increased consumption of goods and services and consequently an increase in waste generated (Hoornweg and Bhada-Tata, 2012a,b).

The MSW is defined as waste collected by the municipality or disposed of at the municipal waste disposal site and includes residential, industrial, institutional, commercial, municipal, and construction and demolition wastes (Hoornweg et al., 2015). Data on SW is generally lacking and, if reported, it is mostly limited to sub-Saharan Africa (Hoornweg and Bhada-Tata, 2012a,b). It has been estimated that Africa contributes about 5% of all waste generated worldwide (Hoornweg and Bhada-Tata, 2012a,b).

The total MSW generated in Africa, in 2012, is estimated to be 125 million ton a year of which 81 million ton is from sub-Saharan Africa (Scarlat et al., 2015). The waste generation rate in Africa is estimated to be 0.65 kg per person per day (varying between 0.09 and 3.0 kg per person per day) and is expected to increase to 0.85 kg per person per day in 2025. This translates into 169,119 ton of waste generated per day in 2012 and 441,840 ton per day in 2025. However, waste generation per capita varies considerably across countries, between cities, and within cities (Hoornweg and Bhada-Tata, 2012a,b). Waste generation is generally lower in rural areas since, on average, residents are usually poor, they purchase less products from stores and therefore generate less packaging waste and are more likely to reuse and recycle items (Hoornweg and Bhada-Tata, 2012a,b).

Africa's rapid growth rate of waste generation (30% between 2012 and 2025) is largely driven by urbanization and increased wealth and is not expected to stabilize before 2100 (Hoornweg et al., 2015). Urbanization in sub-Saharan Africa is expected to result in less dense cities, more akin to the United States than Japan due the availability of land. These types of cities are more likely to be associated with higher volumes of waste being generated (Hoornweg et al., 2015) (Fig. 2.37).

Waste composition is influenced by factors such as culture, economic development, climate, and energy sources (Hoornweg and Bhada-Tata, 2012a,b). Furthermore, waste composition studies are often done as a snapshot in time and do not provide detail on seasonal variability in terms of volumes and composition. Generally, the organic waste percentages in urban waste streams of low- and middle-income countries are high ranging from 40% to 85%. Ash content is generally high in low-income countries where the majority of households are not connected to central electricity supply systems. Paper, plastic, glass, and metal fractions increase in the waste stream of middle- and high-income countries. The average waste composition in sub-Saharan African cities is 57% organic, 9% paper and cardboard, 13% plastics, 4% glass, 4% metal, and 13% others (Fig. 2.38) (Hoornweg and Bhada-Tata, 2012a,b; Scarlat et al., 2015) (Fig. 2.39). The projected changes in composition by country income between 2012 and 2025 are illustrated in Fig. 2.40.

When comparing waste composition data between different cities and even between cities within one country, there is considerable variation, especially in the organic waste

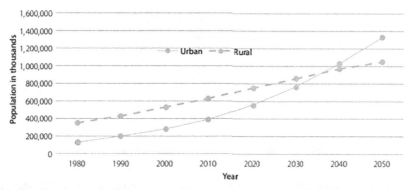

Fig. 2.37 Urban and rural population of African Countries (1980–2050). *(Source: ECS (2014).)*

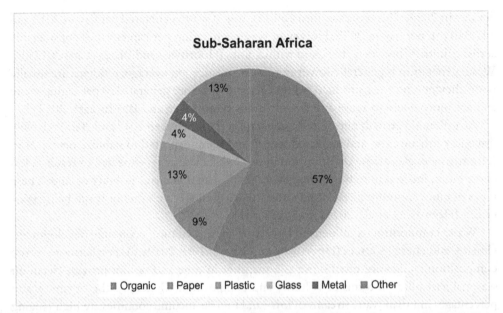

Fig. 2.38 Average MSW composition in Sub-Saharan countries. *(Data from Hoornweg, D., Bhada-Tata, P., 2012. What a Waste: A Global Review of Solid Waste Management. Urban development Series, Knowledge Papers No. 15. World Bank, Washington, DC. © World Bank. https://openknowledge.worldbank.org/handle/10986/17388.)*

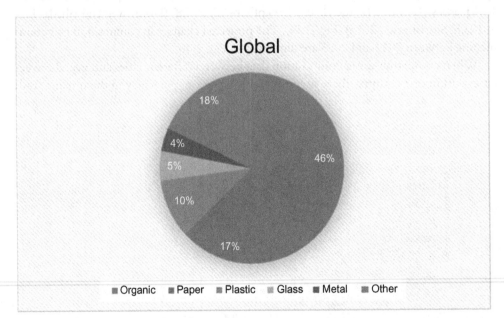

Fig. 2.39 Global MSW composition. *(Data from Hoornweg, D., Bhada-Tata, P., 2012. What a Waste: A Global Review of Solid Waste Management. Urban development Series, Knowledge Papers No. 15. World Bank, Washington, DC. © World Bank. https://openknowledge.worldbank.org/handle/10986/17388.)*

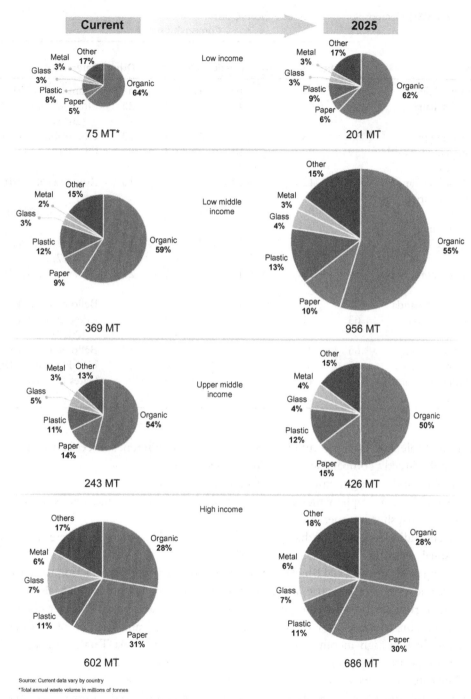

Fig. 2.40 Solid waste composition by income and year. *(Data from Hoornweg, D., Bhada-Tata, P., 2012. What a Waste: A Global Review of Solid Waste Management. Urban development Series, Knowledge Papers No. 15. World Bank, Washington, DC. © World Bank. https://openknowledge.worldbank.org/handle/10986/17388.)*

Table 2.6 MSW Composition for Selected African Cities

City	Composition (%)						
	Organic	Paper/Cardboard	Plastic	Glass	Metal	Other	References
Abudja, Nigeria	56	11	10	4	5	n.a.	Imam et al. (2008)
Accra, Ghana	65	6	4	3	3	20	Oteng-Ababio et al. (2013)
Cairo, Egypt	55	18	8	3	4	12	UN-Habitat (2010)
Cape Town, South Africa	39	20	18	11	7	5	DEADP (2011)
City of Tshwane, South Africa	54	12	10	7	2	16	Komen et al. (2016)
Dar es Salaam, Tanzania	71	9	9	4	3	4	Bello et al. (2016)
Ibadan, Nigeria	70	8	5	2	2	15	Adeyi and Adeyemi (2017)
Johannesburg, South Africa	34	12	19	9	5.0	21	Ayeleru et al. (2016)
Kampala, Uganda	77	8	10	1	0	3	Bello et al. (2016)
Lagos, Nigeria	63	11	4	3	2	20	Adeyi and Adeyemi (2017)
Moshi, Tanzania	65	9	9	3	2	12	Bello et al. (2016)
Nairobi, Kenya	65	6	12	2	1	15	UN-Habitat (2010)
Windhoek, Namibia	47	15	11	14	4	9	Gates Foundation (2012)

fraction (Table 2.6). Data presented in Table 2.6 are from different sources and are not all necessarily directly comparable for the following reasons:

- sampling and sorting methods are not standardized across studies;
- low numbers of samples which may not be representative of the entire city's waste, but rather a snapshot in time;
- not all studies covers more than one season and may therefore not representative of seasonal variation.

According to Hoornweg and Bhada-Tata (2012a,b) the organic waste fraction tends to be highest in low-income countries and lowest in high-income countries. They also report that the organic fraction increase steadily as affluence increase, but at a slower rate than the nonorganic fraction. "Low-income countries have an organic fraction of 64% compared to 28% in high-income countries" (Hoornweg and Bhada-Tata, 2012a,b). The waste composition by country income level is presented in Fig. 2.41.

The status of several African countries are summarized in Table 2.7 in terms of key economic, social, and environmental criteria. Chapter 6 presents detailed information on these countries' waste policies, management practices, and opportunities for MSW to energy.

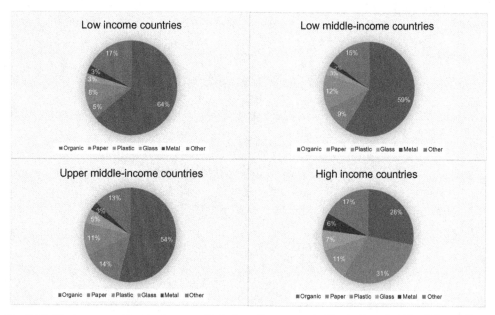

Fig. 2.41 Waste composition by income. *(Data from Hoornweg, D., Bhada-Tata, P., 2012. What a Waste: A Global Review of Solid Waste Management. Urban development Series, Knowledge Papers No. 15. World Bank, Washington, DC. © World Bank. https://openknowledge.worldbank.org/handle/10986/17388.)*

Table 2.7 Key Development Criteria for Selected African Countries

Key Development Criteria for Selected African Countries

	Population	GDP (millions)[a]	GDP/ Capita[a]	Carbon Emissions per Capita (Tons CO_2 Equivalents per Capita)[b]	HDI[c]	Urban Population[d]	MSW per Capita (kg/ capita/ day)[d]	Total MSW (ton per day)[d]
South Africa	54,957,000	761,926	13,403	9.49	0.666	26,720,493	2.00	53,425
Kenya	45,533,000	164,340	3496	1.34	0.555	6,615,510	0.30	2000
Zimbabwe	13,503,000	29,795	2277	1.82	0.516	4,478,555	0.53	2356
Uganda	37,102,000	91,212	2352	0.89	0.493	3,450,140	0.34	1179

[a]Gross domestic product (GDP), based on purchasing power parity. Data from "Report for Selected Country Groups and Subjects (PPP valuation of country GDP)". IMF. Retrieved October 24, 2017.
[b]Carbon emissions for 2013. Data from: Climate Analysis Indicators Tool (CAIT) Version 2.0. (Washington, DC: World Resources Institute, 2014). World Resources Institute. Retrieved 2017-06-12.
[c]Human Development Index, HDI for 2015. Data from "Table 2: Trends in the Human Development Index, 1990–2015."
[d]What a waste. The World Bank. Daniel Hoornweg and Perinaz Bhada-Tata March 2012, No. 15.

CHAPTER THREE

Best Available Technologies (BAT) for WtE in Developing Countries

Suani Teixeira Coelho*, Rocio Diaz-Chavez†
*Research Group on Bioenergy, Institute of Energy and Environment, University of São Paulo, São Paulo, Brazil
†SEI Africa, Stockholm, Sweden

3.1 BIOLOGICAL TREATMENT

This section presents the commercialized technologies in use for the organic matter of municipal solid waste (MSW). It discusses biological process occurring in landfills and biodigestion process occurring when mechanical biological treatment (MBT) is employed. From the biogas produced, technologies are discussed for its upgrading and biomethane production. In addition, technologies commercialized for energy conversion of the biogas and biomethane are analyzed: biogas to energy, biomethane for vehicle use and for grid injection are analyzed. This technology is especially interesting for the developing countries (DCs) since it appears to be the one requiring less capacity building and presenting the lower installation and maintenance costs.

3.1.1 Biological Treatment: Landfills

Vanessa Pecora Garcilasso and Fernando C. de Oliveira
Research Group on Bioenergy, Institute of Energy and Environment, University of São Paulo, São Paulo, Brazil

The generation of solid waste is directly related to the urban population, its standard of living, and consumption habits. The collection, treatment, and adequate disposal of this waste directly affect the quality of life of the population as well as the quality of groundwater and rivers.

Therefore, the final disposal of urban waste becomes one of the serious environmental problems faced by large urban centers around the world and tends to worsen with the increase in the consumption of disposable goods, which increasingly make up the large volumes of garbage generated by the population.

Landfill is one of the forms of final disposal of MSW. It consists of the confinement of waste deposited in the soil, covered with layers of earth isolating it from the environment. Considered environmentally correct, landfills must meet specific operational and

Municipal Solid Waste Energy Conversion in Developing Countries
https://doi.org/10.1016/B978-0-12-813419-1.00003-6

63

environmental standards in order to avoid harm to public health and safety, minimizing negative impacts.

At the stage of preparing the ground for the reception of solid wastes, some preventive measures must be taken to avoid future problems of contamination or lack of stability of the ground, such as soil sealing and fluid drainage system, among others, which also contribute to the optimization of the collection of the generated biogas.

Soil sealing must follow criteria that will depend on its own characteristics and the characteristics of the climate. Generally, it is made by layers of clay and geo-membranes of high-density polyethylene (HDPE), which is used to prevent infiltration of percolated liquids (slurry) into the soil, which contaminates the groundwater (Inovageo, 2017). Then, a new layer of clay is placed (Sansuy, 2017).

From this preparation, the garbage is deposited on the ground, compressed, and then covered with layers of soil from the site itself, isolating the waste from the environment. Each layer of the landfill is about 5 m high, consisting of compacted trash, soil, clay, and geo-membrane. The compactness of the waste allows a higher density to be obtained, increasing the gas production per unit volume, while the appropriate coverage prevents the entry of water from runoff, in addition to preventing the entry of oxygen and the escape of biogas to the atmosphere. The infiltration of oxygen would slowdown the process of anaerobic decomposition—the phase at which methane is produced (Borba, 2006).

During the process of anaerobic digestion of the organic matter, present in the MSW, there is a formation of two environmental polluting vectors: slurry—a dark and nauseating-odor liquid—and biogas.

The slurry is caught by means of horizontal pipes, implemented during the grounding of the MSW, and drained to treatment or retention tanks, where it is stored and subsequently transported to a treatment plant. If the slurry is not effectively collected, it may lead to increased moisture of the waste and to pollution of water resources. Circumventing this problem requires the slurry to be properly collected and the soil properly waterproofed, as mentioned earlier.

In order to avoid excessive rainwater, pipes are placed around the landfill cells with the function of intercepting and diverting the flow of this water to ponds and for further use of it, avoiding its infiltration into the grounded MSW.

For the collection of biogas in a landfill, suction pipes are placed horizontally when the waste is still being deposited, which allows biogas to be collected from the beginning of its production (Willumsen, 2001), while vertical pipes are placed at specific points in the landfill area—and connected to horizontal pipes—to enable biogas extraction (Fig. 3.1) (Cenbio, 2006). The term "biogas" is here defined as the gaseous product of the anaerobic digestion of organic matter from MSW, regardless of whether it is obtained from a sanitary landfill or from biodigesters.

The transportation pipeline is connected to each vertical pipe in order to send the biogas to the extraction system (blowers) and, subsequently, to the system burning in

Fig. 3.1 Biogas collection pipes. *(Data from Cenbio, 2006. Available online in 25 June 2019. http://www.iee.usp.br/gbio/?q=biogás.)*

flare, or to other end uses, such as energy. It is worth pointing out that the biogas extraction process at landfill sites allows the recovery of 40%–60% of the total biogas produced.

According to the licenses obtained for the installation and operation of the landfill, the area intended for the grounding of the garbage, denominated cell, is closed when the amount of deposited waste reaches the maximum permissible height. The landfill may contain several cells, depending on the characteristics of the land and the amount of residue deposited there.

After the closure of the landfill, the tendency is that no more biogas production will occur, since the organic matter of the MSW responsible for the biogas production is zero, as there is no longer the deposition of waste in the landfill. However, the organic matter that had already been grounded will continue to produce biogas, which must be captured for later burning in flare or to be used in other energy purposes.

The area destined for the disposal of the garbage will also be unable to be used for civil constructions after the closure of the landfill activities. Generally, these areas are used for the construction of green areas, such as parks, in which gas generation will continue to be monitored.

If the biogas generated in landfills is not captured or properly controlled, it represents risks to the environment and the population, as it may migrate to the areas near the landfill or emanate from its surface.

Used as a device to burn biogas, flare is considered a component of each energy recovery option, as it may be necessary during the early stages of the process and maintenance of the system. It can also be used for the biogas-burning surplus between system upgrades (Muylaert, 2000).

After collecting the biogas produced in a landfill, and before its use in the process of energy conversion, it must be treated for the removal of moisture, particulates, and impurities in general, so that it can be harnessed according to the energetic needs of the local population, such as electric and thermal power generation.

The energy from the MSW gains importance under the new energy generation policies from biomass and other renewable sources since they can reduce the consumption of fossil fuels. In general, landfills have a high capacity to generate electricity from biogas, because the energy generated in this way can decrease the dealer's overload and reduce the emission of greenhouse gases (GHGs), since methane—the main constituent of biogas—is transformed into carbon dioxide (CO_2), whose global warming potential is 21 times lower. In addition, there is the possibility that the landfill market the surplus electricity with the local utility company.

However, it is important to mention about the huge number of existing dumps in Brazil, where part of the MSW is still inadequately destined or deposited in controlled landfills (ABRELPE, 2016). Thus, the National Policy of Solid Waste (PNRS, Brazilian acronym) aims to end such dumps—a practice banned in the country for a long time, but not yet fulfilled—and to prevent the disposal of *in natura* residues in landfills.

3.1.2 Biological Treatment: Biodigestion

Vanessa Pecora Garcilasso and Fernando C. de Oliveira

Research Group on Bioenergy, Institute of Energy and Environment, University of São Paulo, São Paulo, Brazil

Put simply, the anaerobic digestion is the process of decomposition of organic matter present in the substrate to be degraded in the absence of oxygen. This process occurs in four stages: hydrolysis, acidogenesis, acetogenesis, and methanogenesis (GIZ, 2010). It can be used as a stand-alone option and also as part of the MBT. MBT is the combination of recycling, organic matter biodigestion, and a thermal treatment process for the nonorganic waste that was not recycled.

Hydrolysis is the first stage in the process of anaerobic degradation in which hydrolytic bacteria release enzymes that breakdown complex organic compounds—such as carbohydrates, proteins, and lipids—into less complex substances—such as amino acids, sugars, and fatty acids—in order to be processed by the cells.

The next phase is acidogenesis, in which the acidogenic fermentative bacteria enable the intermediate compounds to be decomposed into short-chain fatty acids (acetic, propionic, and butyric acids), in addition to carbon dioxide and hydrogen. At this stage, the formation of small amounts of lactic acid and alcohols also occurs.

The third phase, acetogenesis, is responsible for the oxidation of the products generated in the previous phase (acidogenesis), yielding substrates suitable for methanogenic microorganisms, forerunners of biogas. The products generated in this phase are acetic acid, hydrogen, and CO_2.

In the fourth and last phase, strictly anaerobic bacteria, called methanogenic bacteria, which produce methane and CO_2, consume the organic compounds formed in acetogenesis phase.

The main parameters that influence the process of anaerobic digestion are:

- *Absence of oxygen*: methanogenic bacteria are strictly anaerobic. The decomposition of organic matter in the presence of oxygen produces only CO_2.
- *Substrate composition*: the characterization of organic matter is of great importance in determining or predicting the efficiency of the anaerobic digestion process. The higher the percentage of organic matter presents in the waste to be treated, the greater the biogas production potential.
- *Temperature*: sudden temperature changes cause imbalance in the cultures involved in the anaerobic digestion process, especially in the methanogenic bacteria. The microorganisms must be adapted to the temperature range of work, allowing to classify the processes in:
  - Psychophilic, when psychophilic bacteria are used, operating in a range below 20°C.
  - Mesophilic, when mesophilic bacteria are used, operating in a range of 20–45°C.
  - Thermophilic, when thermophilic bacteria are used, operating in a range of 45–60°C.

Below 10°C, the process is generally interrupted as the gas production increases with the rise in the temperature. Above 65°C, the enzymes are destroyed by heat. In addition, the pH of the medium, acidity, and alkalinity, are important factors for the anaerobic digestion process. The pH of the medium should be neutral and maintained between 6 and 8, being considered optimal between 7 and 7.2. The pH control is performed as a function of the accumulation of bicarbonate and the concentration of ionized volatile acids, nitrogen in the form of ammonia, and CO_2.

Biogas can be obtained from various substrates, such as municipal waste (solid and liquid), agricultural residues, waste from food and beverage industries, and agro-industrial waste.

The large volume of waste generated in these activities presents a high polluting load that leads to environmental, social, and economic damages. Thus, the correct treatment of these residues in anaerobic systems, besides the ability to depollute, allows the energetic use of biogas, which can accelerate the amortization of the costs of the systems used.

In the case of landfills, once the MSW is grounded, the organic matter present begins to decompose, generating biogas. The biogas production curve in a landfill is increasing until the end of its activities, that is, until it stops receiving MSW. After its closure, the biogas curve begins to fall, because there will be only biogas production of the organic matter that is already grounded. Depending on the landfill conditions and other factors, such as waste composition, climatic factors, among others, there will still be biogas production for more than 10 years.

3.1.3 Biogas to Energy

Marilin Mariano dos Santos and Caio Luca Joppert
Research Group on Bioenergy, Institute of Energy and Environment, University of São Paulo, São Paulo, Brazil

Biogas produced from MSW is a mixture of methane and carbon dioxide. It also contains small amounts of water, hydrogen sulfide, ammonia, heavy metals, halogens, and small particles. Due to the presence of contaminants, the end use of biogas requires some type of pretreatment, and the type of treatment implemented is directly related to the end use and to the contaminants present. Hence, depending on the end use of biogas, it must comply with the specific quality requirements. Table 3.1 (adapted from Wellinger et al., 2013) shows the average composition of landfill biogas and biogas from anaerobic digestion of MSW in biodigesters.

Regarding end uses of biogas from MSW, its conversion to energy can occur several ways. It can be used in the same way as natural gas or others combustible gases like liquefied petroleum gas (LPG).

Typical end uses for biogas from MSW include hot water, lighting, and electricity generation. Biogas can also be purified to biomethane and then be used in the transport sector or injected into the natural gas grid, situation in which biomethane acts as a substitute of natural gas. In this case, biomethane is commonly named renewable natural gas (RNG). Fig. 3.2 (Patrizio et al., 2015; Khan et al., 2017) shows a flow diagram in which biogas end its uses separates into to four categories: electricity, heat, transport, and RNG.

The most common end use of biogas is the production of heat and electricity. In most cases, biogas is converted into electricity and heat using Otto cycle or Diesel cycle engines adapted for biogas, turbines, and full cells. However, before biogas can be used, it must be conditioned. Table 3.2 adapted from Frandsen et al. (2011) presents the technologies utilized to convert biogas to energy.

It stands out that the end use of biogas depends on local structural, such as tax systems, subsidies, investment programs, availability of natural gas networks, etc.

3.1.3.1 Technologies for Power Production From Biogas

Biogas could be an economical fuel for generation of electricity and heat. Technologies like internal combustion engines (ICEs), gas turbines, micro turbines, Stirling engines, and fuel cells have been used to generate electricity using biogas as fuel.

This chapter will discuss the main technologies utilized in the end uses of biogas, such as micro turbines and engines.

Table 3.1 Typical Components in Biogas From Landfills and From Anaerobic Digesters

Compound	Unit	Landfill Gas	AD Gas
Methane	% vol.	30–65	50–70
Carbon dioxide	% vol.	25–47	30–50
Nitrogen	% vol.	0–17	2–6
Oxygen	% vol.	0–3	0–5
Hydrogen sulfide	mg/m^3	0–1000	100–10,000
Ammonia	mg/m^3	0–5	0–100
Siloxanes	mg/m^3	0–50	0–50

Data from Wellinger, A., Murphy, J.P., Baxter, D., 2013 The Biogas Handbook: Science, Production and Applications. Elsevier, Cambridge.

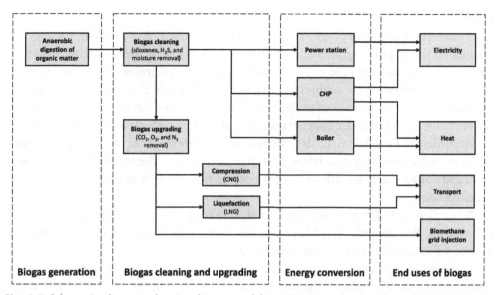

Fig. 3.2 Schematic diagram showing biogas and biomethane end uses. *(Data from Patrizio, P., Leduc, S., Chinese, D., Dotzauer, E., Kraxner, F., 2015. Biomethane as transport fuel – a comparison with other biogas utilization pathways in northern Italy. Appl. Energy 157, 25–34; Khan, I.U., Othman, M.H.D., Hashim, H., Matsuura, T., Ismail, A.F., Rezaei-Dashtarzhandi, M., Azelee, I., 2017. Biogas as a renewable energy fuel – a review of biogas upgrading, utilization and storage. Energy Convers. Manage. 150, 277–294).*

Table 3.2 Overview of Technologies for Utilization and Upgrading Biogas

End Use		Technology
Power		Micro turbine
		Engines
Heat		Boiler
CHP—combined heat and power		Cycle diesel engines
		Cycle Otto engines
		Stirling engines
		Micro turbine
Injection into the natural gas grid	Biogas to biomethane (biogas upgrading)	PSA—pressure swing adsorption
		Absorption
		Water scrubbing
		Physical scrubbing
Vehicular		Chemical scrubbing
		Membrane separation
		Cryogenic processes

Data from Frandsen, T.Q., Rodhe, L., Baky, A., Edström, M., Sipilä, I.K., Petersen, S.L., Tybirk, K., 2011. Best available technologies for pig manure biogas plants in the Baltic Sea Region. In: Baltic Sea 2020, Stockholm.

Engines

The most common technology for power generation is ICEs, probably because gas engines do not require high gas quality; it is only necessary to condensate water vapors and reduces the concentration of H_2S below 250 ppm. These treatments are necessary to avoid the corrosion of gas engines in power plants. According to traditional manufacturers, the electrical efficiency of gas-powered engines with biogas is around 35% without cogeneration; if we consider cogeneration, the efficiency is about 80% (USEPA, 2017).

According to Dudek et al. (2010), the reason ICEs represent the most commonly used technology for electricity generation from biogas is because the technology is predominant and consolidated. In addition, the economic risks are very low when compared to other technologies.

Regardless of ICE representing the most used technology for electricity generation from biogas, they are only available for powers above 100 kW and require biogas to have at least 45% methane content (Hakawati et al., 2017).

In regard of the capital cost for gas engines, Table 3.3 (adapted from AMPC, 2017) presents the estimated values of equipment and total costs of the project related to the nominal capacity of the engines. These figures do not include the costs to obtain and treat biogas. Is important to note that for assembling a biogas plant, it is common to use more than one motor-generator set in parallel.

Micro Turbines

A power plant that uses gas turbines operates based on a thermodynamic Brayton Cycle, independently of the type of the gas. A gas turbine that uses biogas is very similar to a natural gas turbine, except for the number of valves and fuel regulator injectors, which is double.

Table 3.3 Estimated Capital Cost for Typical Gas Engine Generators With Heat Recovery

Source Data	Engine Output (kWe @415V)	Heat Output (kWth @900°C)	Engine Plus Heat Recovery and Commission (US$)	Equipment and Commission (US$/kWe)	Cost Inc. Install & Project Costs (US$/kWe)[a]
Clarke	300	368	583,430	1945	
	847	977	839,828	991	
	1067	1234	915,428	857	
US	633			912	2432
EPA	1121			887	2175

[a]Values for the US. The costs for a complete project is significantly greater and includes allowances for interconnections, material, and labor (duct work, piping, wiring, etc.), project management, engineering, contingency (approx. 5% of equipment cost) and project financing.
Data from AMPC, 2017. Investigation Into Modular Micro-Turbine Cogenerators & Organic Rankine Cycle Cogeneration Systems for Abattoirs. Final Report. Australian Meat Processor Corporation, Sydney Australian, p. 39. Project Code 2016.1002.

Commercial biogas turbines have capacities that range from 25 to 1000 kW. Due to low nominal power outputs, they are called micro turbine. Its efficiency is comparable to low-power Otto cycle engines, around 30% (without considering cogeneration) (Capstone Corporation, 2017).

It must be noticed that the biogas supplied to the turbine must be at pressures in the order of 5.5 bar. Thus, it is necessary to use a compressor, which consumes a significant part of the energy generated and affects the overall efficiency of the system (USEPA, 2017).

Manufacturers have established a fuel specification that requires $<0.03 \, mg/m^3$ of siloxanes, as prolonged exposure to biogas with high siloxanes content results in a progressive loss of performance. This is mainly due to the accumulation of silicate in the combustion chamber and in the regenerator (Dudek et al., 2010).

Some manufacturers have not confirmed siloxanes contamination, but maintain an official fuel restriction of 10 ppbv of siloxanes (Wheles and Pierce, 2004; McBean, 2008). Regarding the hydrogen sulfide contamination, some manufacturers claim to accept high levels of H_2S (up to 70,000 ppm), but others can operate with concentrations only up to 5000 ppm (Capstone Corporation, 2017; Clarke, 2017).

Considering the biogas cleaning system to be burned in turbines, mainly regarding the removal of siloxanes, particulate material (PM), and condensable, the cost of a power generation project can be higher than the one observed in the use of gas engines.

Micro turbines can work with biogas with methane content as low as 30% and have electrical efficiencies from 22% to 30%. These figures can reach values of 63%–70% for projects with combined heat and power (CHP) (Dudek et al., 2010). In addition, micro turbines have low atmospheric emissions and allow recovering the heat from the exhaust gases as low-pressure steam, an important environmental benefit (Persson et al., 2006).

Regarding the capital cost for micro turbines, Table 3.4 (adapted from AMPC, 2017) presents the estimated values of equipment and costs of the project related to the nominal

Table 3.4 Estimated Capital Cost for Typical Micro Turbine With Heat Recovery

Nominal Installed Power	kWe	200	1000
Generator package	$US	359,300	1,188,600
Heat recovery	$US	0	275,000
Gas compression	$US	42,600	164,000
Equipment cost	$US	401,900	1,627,600
Cost Inc. install and project costs	US$[a]	196,600	747,300
Total installed cost	US$	598,500	2,374,900

[a]The costs for a complete project is significantly greater, and includes allowances for interconnections, material and labor (duct work, piping, wiring, etc.), project management, engineering, contingency (approx. 5% of equipment cost) and project financing.
Data from AMPC, 2017. Investigation Into Modular Micro-Turbine Cogenerators & Organic Rankine Cycle Cogeneration Systems for Abattoirs. Final Report. Australian Meat Processor Corporation, Sydney Australian, p. 39. Project Code 2016.1002.

capacity of the micro turbine. These figures do not include the costs to obtain and treat biogas.

Finally, yet importantly, it must be mentioned that one of the main advantages of micro turbines in comparision to engines is the low emission of pollutants.

Combined Heat and Power

The combined generation of heat and power (CHP) is based on recovering the waste thermal energy content in the exhaust or flue gases of thermal machines, which is often rejected to the environment. This waste heat can be used for space heating and cooling, water heating, and process heat.

Biogas CHP projects can use thermal machines such as ICEs, gas turbines, or micro turbines. Technologies less common to electricity generation and heat include boiler/steam turbine applications and fuel cells.

CHP process increases the overall efficiency of the system from 25%–55% to 60%–90%, depending on the equipment used and the application. In a modern CHP system, about 10%–15% of the energy of the biogas is wasted. The electrical efficiency is low: from $1.0\,Nm^3$ of biogas, only 2.4 KWh of electric energy can be produced. In a typical CHP plant, on average, 30% of the input energy is transformed into electricity, 50% into heat, 5% is lost by radiation, and 15% lost in the exhaust gases (NNFCC, 2010).

Investment costs for biogas plants range from 1660 to 2395 \$US/kWh if power plant generates only electricity and from 1840 to 2580 \$US/kWh if heat is a secondary product. Similarly, as with other energy generation equipment, the pretreatment of the biogas is required in order to remove of H_2S, moisture, and siloxanes before use (Niemczewska, 2012).

3.1.3.2 Biogas Upgrading, Biomethane Vehicular Use, and Biomethane Injection Into the Grid

Biogas cleaning and upgrading is necessary for uses like natural gas substitution (injection into the natural gas grid) or vehicle fueling. In both cases, the biogas composition must compatible with natural gas composition in terms of proprieties. It is important to note that, to be injected into the natural gas distribution network and used as a vehicular fuel, biomethane must comply with the legal requirements for fuel quality.

The regulatory entities determine the minimum requirements that biomethane needs to comply with for natural gas network injection or to be used as a vehicle fuel. Regarding biomethane composition and properties, some of the parameters regulated are: minimum methane concentration; maximum concentration of ethane, propane, butane, oxygen, carbon dioxide, nitrogen and sulfidric gas, total sulfur, siloxanes, ammonia as well as dew point of water and hydrocarbons, Wobbe index, density, lower heating value (LHV), among others.

In Europe, the standard EN 16723-1 (EU, 2016) regulates specifications of biomethane for injection in the natural gas network and the standard EN Draft

16723-2 (EU, 2016) regulates automotive fuel specifications. Some countries have their own regulations according to their particularities. For instance, Brazil has the technical regulation ANP 685 of June 29, 2017 (ANP, 2017), which establishes the rules for the approval of quality control and the specification of biomethane from landfills and sewage treatment plants for vehicular use and residential and industrial installations to be distributed throughout the national territory. Table 3.5 (Khan et al., 2017) exemplifies the magnitude of the regulated parameters. They were determinated aiming to prevent unwished effects during the end uses, to raise the volumetric energy content and lower risks to public health.

To achieve these values, it is necessary to clean and upgrade biogas. Biogas upgrading has the objective of reducing carbon dioxide (CO_2) concentration; biogas cleaning aims to reduce the contaminants content, such as ammonia (NH_3), hydrogen sulfide (H_2S), siloxanes, moisture, etc.

Fig. 3.3 (Wellinger et al., 2013) shows the technological routes for cleaning and upgrading biogas and Table 3.6 (adapted from Ryckebosch et al., 2008) presents the technologies for CO_2, H_2S, siloxanes, and moisture reduction.

According to Khan et al. (2017), the amount of energy required to upgrade raw biogas is another important factor when selecting a technology route. The water scrubbing and organic physical scrubbing technologies are cheaper and the energy consumption is

Table 3.5 Regulated Values for Biomethane in Some Countries

	Unit	Sweden	France	Switzerland	German	Netherlands	Austria
CH_4	% vol.	≥ 97	≥ 86	≥ 96	≥ 96	≥ 85	≥ 96
CO_2	% vol.	≤ 3	≤ 2.5	≤ 6	≤ 6	≤ 6	≤ 3
O_2	% vol.	≤ 1	≤ 0.01	≤ 0.5	≤ 0.5	≤ 0.5	≤ 0.5
H_2	% vol.	≤ 0.5	≤ 6	≤ 4	≤ 5	≤ 0.5	≤ 4
CO	% vol.	–	≤ 2	–	–	≤ 1	–
H_2S	mg/Nm^3	≤ 10	≤ 5	≤ 5	≤ 5	≤ 5	≤ 5
Total sulfur	mg/Nm^3	≤ 23	≤ 30	≤ 30	≤ 30	≤ 16.5	≤ 10
NH_3	mg/Nm^3	≤ 20	≤ 3	≤ 20	–	≤ 3	0
H_2O	mg/Nm^3	≤ 3	–	–	–	–	–
Water dew point	°C	≤ -5	≤ -5 P_{max}	–	Soil temp	≤ -8, 70 bar	≤ -8, 40 bar
Heavy metals	mg/Nm^3	–	≤ 1	≤ 5	≤ 5	–	–
Siloxanes	mg/Nm^3	–	–	–	–	≤ 5	≤ 10
Halogens	mg/Nm^3	–	≤ 1 (Cl) ≤ 10 (F)	≤ 1	0	$\leq 50/25$ (Cl/F)	0
Mercaptans	mg/Nm^3	–	≤ 6	≤ 5	≤ 15	≤ 6	≤ 6

Data from Khan, I.U., Othman, M.H.D., Hashim, H., Matsuura, T., Ismail, A.F., Rezaei-Dashtarzhandi, M., Azelee, I., 2017. Biogas as a renewable energy fuel – a review of biogas upgrading, utilization and storage. Energy Convers. Manage. 150, 277–294.

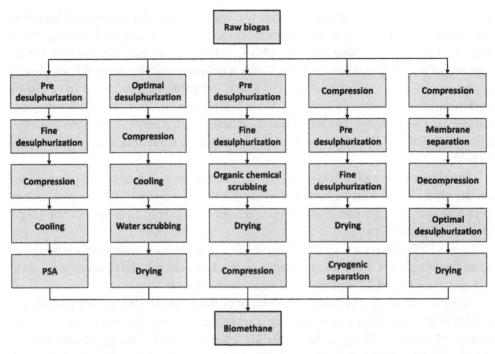

Fig. 3.3 Technological routes for cleaning and upgrading biogas. *(Data from Wellinger, A., Murphy, J.P., Baxter, D., 2013. The Biogas Handbook: Science, Production and Applications. Elsevier, Cambridge.)*

Table 3.6 Technologies for Reduction of Concentration of CO_2, H_2S, Siloxanes and Humidity

Compound	Separation Method	Usual Process
CO_2	Adsorption	Pressure swing adsorption using zeolite beds
	Physical absorption	Pressurized water scrubbing
	Chemical absorption	Monoethanolamine (MEA) scrubbing
	Membrane separation	Polymer membrane gas separation (dry)
		Membrane gas separation (wet)
	Cryogenic process	Low temperature separation process
H_2S	Chemical adsorption	Adsorption using iron sponge or zinc oxide beds
	Physical adsorption	Adsorption using activated carbon beds
	Chemical absorption	Amine or water scrubbing
	Chemical absorption	NaOH scrubbing
	Biological	Biological removal on a filter bed
Siloxane	Physical adsorption	Adsorption using activated carbon beds
Humidity	Cooling process	Condensation

Data from Ryckebosch, E., Drouilon, M., Vervaeren, H., 2008. Techniques for transformation of biogas to biomethane. Biomass Bioenergy 35, 1633–1645.

0.20–0.43 and 0.4–0.5 kWh/Nm^3, respectively. On the other hand, the pressure swing adsorption (PSA) has an energy consumption range of 0.24–0.60 kWh/Nm^3. Yet, according to Khan et al. (2017), the cost of capital investment considering the various upgrading technologies is from 0.49 to 0.15 US\$/Nm^3 of biogas. The economic impact of upgrading can be evaluated considering the quantity of biogas processed and the technology used.

The analysis of the end uses of biogas shows that the use for electric generation is still the main end use of biogas. The substitution of natural gas by biomethane and its vehicular use are the most often found in the developed countries, such as in Sweden, Germany, Denmark, Italy, and others. As an example of the primacy of use for electric generation, in the year 2016, there were 17,632 biogas plants in Europe with an installed capacity of 9985 MWe. Among the countries of Europe, the United Kingdom is the one that produces the most electricity from landfill gas (LFG) −7290 GWh. Regarding biomethane the production in Europe, the volume produced in 2015 was 1.23 billion Nm^3. The countries that saw the most significant biomethane development in 2016 were Germany (+900 GWh), France (+133 GWh), and Sweden (+78 GWh) (EBA, 2017).

The supremacy of the use of biogas to electricity generate in relation to biomethane uses, even in the developed countries, is justified by the complexity of the processes involved for biomethane production and the costs involved.

3.2 THERMOCHEMICAL TREATMENT

Suani Teixeira Coelho and Rocio Diaz-Chavez

Research Group on Bioenergy, Institute of Energy and Environment, University of São Paulo, São Paulo, Brazil
SEI Africa, Stockholm, Sweden

This section analyses the existing commercialized waste to energy (WtE) technologies using thermochemical treatment, in particular MSW-incineration and gasification processes. General information on the technological aspects and a discussion on economic, environmental, and social aspects are presented for each one. These technologies are the most used in the developed countries where existing legislation do not allow the disposal of *in natura* waste in landfills. MSW must pass through the recycling and recovery processes, before the WtE processes and the residues are disposed in landfills.

However, for the DCs, there are still several difficulties, mainly economic, as discussed in this chapter.

In fact, as discussed in Diaz-Chavez and Coelho (2017), "waste to energy may need more incentives for companies managing waste but also for municipalities to promote more effective recycling systems and more incentives to society to reduce consumerism and waste. Waste should not be considered an environmental problem but as an opportunity for energy generation and for the circular economy. Shifting this thinking will move it to another paradigm where different stakeholders should be involved from the governance point of view."

Table 3.7 Appropriate Technologies for Small Municipalities in Brazil

Amount of MSW	Potential Electricity Generation
1200 ton/day (large municipalities)	20 MW (incineration)
60 ton/day (municipality of 60,000 people)	1 MW
5 ton/day (municipality of 5000 people)	75 kW approx. (gasification)

It is worth to note that these WtE thermochemical technologies have different capacities and are suitable for different situations. Incineration systems must have a minimum output of 5–10 MW (which corresponds to large municipalities, with more than 600 people); below this figure, the most suitable are the gasification systems, such as in the case of 60,000 people, with a potential for 1 MW of installed power.

Table 3.7 illustrates the different energy recovery options for different sizes of municipalities in the country, based on the Brazilian experience (Coelho, lecture notes).

Thermochemical processes are deemed as promising and viable solutions in the DCs because of their productivity and compatibility with existing infrastructure. They are in development and still face challenges including technical issues, costs, and social acceptance, but have opportunity to be better implemented. Some of these technologies are applied also to biofuels that can be used for heat and power or as feedstock for the chemical industry (Rago et al., 2012).

3.2.1 Thermochemical Treatment: Incineration

Fábio Rubens Soares and Suani Teixeira Coelho

Research Group on Bioenergy, Institute of Energy and Environment, University of São Paulo, São Paulo, Brazil

3.2.1.1 Incineration Basic Concepts

MSW incineration, which is the process of MSW combustion, corresponds to the WtE process most used in industrialized countries. In Japan, almost 100% of MSW is incinerated, around 80% in Switzerland, Sweden, and Germany, and 30 million tons per year of incinerated waste in the United States. In Paris, 100% of MSW is incinerated with use of thermal energy to heat about 70,000 apartments. In the recent years, several incineration plants have been installed also in East Asian countries (IEA, 2010).

Table 3.8 shows data from some WtE plants installed around the world with incineration technology.

In Brazil, currently, incineration is used only to solve the final disposition of hazardous solid waste, however, without taking advantage of thermal energy obtained. The first MSW incineration plant in Brazil is the Barueri Energy Recovery Unit (ERU), in Barueri municipality, Sao Paulo State.

In general, worldwide these plants use MSW incineration to produce steam to be fed into a steam turbine coupled to a generator to produce electricity or to be used directly in a CHP (or cogeneration) system for an industrial process or for district heating.

Table 3.8 Installed Capacity of Some Incineration Plants in Operation

Plant Location	Treatment Capacity (ton/day)	Energy Production (MW)	Energy Potential (MWh/ton)
Tsurumi, Japan	600	12	0.48
Tomida, Nagoya, Japan	450	6	0.32
Dickerson, Maryland, United States	1800	63	0.85
Alexandria, Virginia, United States	975	22	0.54
Isvag, Antwerp, Belgium	440	14	0.76
Savannah, United States	690	12	0.42
Izmit, Turkey	96	4	1.00
UIOM Emmenspitz, Switzerland	720	10	0.33
Wells, Austria	190	7	0.88
AVG			0.62

Data from Menezes, R.A.A., Gerlach, J.L., Menezes, M.A., 2000. Estágio Atual da Incineração no Brasil. In: Seminário Nacional de Resíduos Sólidos e Limpeza Pública, 7, Curitiba, 3 a 7 de abril de 2000. Anais...ABL – Associação Brasileira de Limpeza Publica, Curitiba.

Table 3.9 LHV of Some Materials Found in MSW (Average Composition)

Material	kcal/kg
Plastic	6300
Rubber	6780
Leather	3630
Textiles	3480
Wood	2520
Food	1310
Paper	4030

Data from EPE [Empresa de Pesquisa Energética], 2007. Plano Nacional de Energia 2030. EPE, Rio de Janeiro. Disponível em http://www.epe.gov.br. (Acesso em: 23 out. 2013).

The electricity generation process through MSW incineration is a conventional Rankine thermoelectric power plant, with the necessary exhaust gases cleaning system. The generation capacity depends directly on the cycle thermodynamic efficiency and on the MSW-LHV.

The LHV, usually expressed in kJ/kg, is calculated from the expression formulated by Themelis (2013) based on the statistics collected in field research:

$$LHV = \left[18,500 \times Y_{fuel} - 2.636 \times Y_{H_2O} - 628 \times Y_{glass} - 544 \times Y_{metal}\right]/4.185$$

The variables Y_{fuel}, Y_{H_2O}, Y_{glass}, and Y_{metal} represent the proportion of each component in 1.0 kg of MSW to be converted. Table 3.9 presents the LHV of the components usually found in MSW (despite the variation among countries) and evaluate that MSW with a high fraction of organic material tend to have lower heating value.

Although the classification according to the LHV is not definitive to define the best MSW destination, Themelis (2013) considers that:

- for LHV < 7000 kJ/kg, incineration is not technically feasible (besides technical difficulties, it also requires the addition of auxiliary fuel);
- for 7000 kJ/kg < LHV < 8400 kJ/kg, incineration technical feasibility depends on some type of pretreatment to increases LHV; and
- for LHV > 8400 kJ/kg, incineration is technically feasible.

Fig. 3.4 shows a schematic diagram of MSW incineration.

Incineration plants can generate between 400 and 700 kWh/ton of MSW. The most used technology is known as "mass burning" (incineration process where all residue is burned, without any processing) (Themelis, 2013). There are still no MSW incineration plants with energy recovery in commercial scale in Brazil, although there is already one project underway in Barueri, in Sao Paulo State.

The efficiency of MSW incineration plants without cogeneration is relatively low, between 20% and 25%, due to the restriction of not operating at higher temperatures. In fact, the current state-of-the-art technology corresponds to the MSW burning at temperatures below 450°C in order to avoid equipment corrosion (Tolmasquim, 2003).

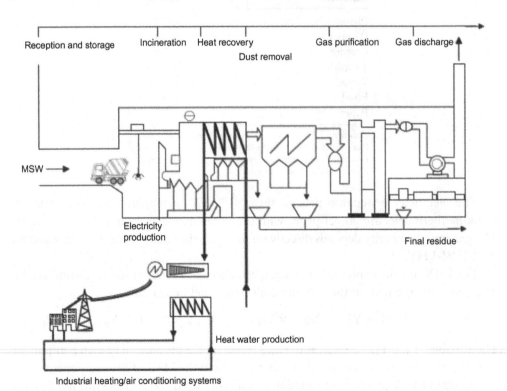

Fig. 3.4 Energy use in the incineration process. *(Data from Menezes, R.A.A., Gerlach, J.L., Menezes, M.A., 2000. Estágio Atual da Incineração no Brasil. In: Seminário Nacional de Resíduos Sólidos e Limpeza Pública, 7, Curitiba, 3 a 7 de abril de 2000. Anais...ABL – Associação Brasileira de Limpeza Publica, Curitiba.)*

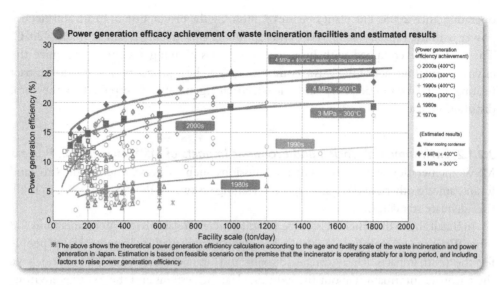

Fig. 3.5 Efficiency of MSW incineration facilities in Japan. *(Data from Japan Environment Ministry, 2015. Municipal Waste Incineration Technology, Safe and Sound Waste Incineration and High-Efficiency Power Generation. Waste Recovery Commission, Japan Environment Ministry, Tokyo, Japan.)*

The priority factor in the installation of a waste incineration plants is the pollution control (elimination of heavy metals, dioxins, and furans, as discussed ahead), which results in a significant improvement of facilities. Priority is not the efficiency of energy recovery process.

It is well known that increasing temperature and steam pressure for electricity generation results in high efficiency. However, acid compounds in the exhaust gases at high temperature produce corrosion in the steam superheating system. In the recent years, research has advanced developing long service heat transfer tubes that show resistance to high-temperature corrosion. Many plants built recently present highly efficient electricity generation facilities with longer operational lifetime (Fig. 3.5).

The reduction of MSW volume to be disposed in a landfill is one of the main advantages of incineration. In addition, the European Union (EU) legislation does not allow the disposal of *in natura* waste in landfills; the same is forecasted to happen in Brazil with the National Policy on Solid Residues (NPSR), hopefully to be applied in a near future. A similar situation is found in the other DCs.

Incineration reduces the volume of waste deposited between 85% and 90% of the original volume and does not prevent recyclable metals recovery. There is also the possibility that ashes from the incinerator are used as raw material for the cement production, if they comply with the local environmental legislation (Oliveira, 2003).

Sweden is a good example for incineration process in use, since it has several MSW incineration plants.

The Boras Thermal Treatment Unit is a MSW incineration plant with energy recovery through CHP process. This plant is located very close to the city and has been operating since 2005. The plant operates combining the production of steam and electric power with the capacity of 45 MWe and four boilers, producing steam at 48 bar, 420°C. The boilers are equipped with high-technology gas cleaning system to minimize the environmental impact of the waste combustion. The heat production is used for district heating in Boras municipality and the electricity generated is supplied to the local distribution grid.

3.2.1.2 Environmental, Social, and Economic Impacts of MSW Incineration Process

MSW incineration generates pollutant emissions such as heavy metals, dioxins, and furans, among others, and requires special cleaning system for the exhaust gases, to fulfill the adequate standards.

In Brazil and in the other DCs, there are still concerns related to MSW incineration and other WtE process, mainly due to the lack of adequate information. In general, local people fear the health impacts of heavy metals, dioxins, and furans emissions since they do not have the adequate information about the existing cleaning systems and the existing environmental legislation. A more detailed overview about energy supplying coming from waste in Brazil is included in Coelho et al. (2017). There are other papers including a discussion about the Waste Governance and Sustainability in Coelho et al. (2016, 2017).

In addition, ashes produced in incineration also contain pollutants. Because of the cleaning systems required to comply with the environmental regulation, the World Bank (Rand et al., 2000) analyzes that incineration plants present high investment, operation, and maintenance costs, and this is the reason why there are few plants in the DCs.

The EU has quite strict standards for pollutant emissions in incineration plants[1] and existing measurements show emissions quite lower than the limits established.

As the design of a WtE plant must meet the standards for pollutant emissions, the investment must take into account these aspects. For example, in Sao Paulo State, in 2008 CETESB (Sao Paulo State Environmental Agency) issued a regulation (SMA 079, November 2009[2]) following the same limits of the EU for any plant in the state. The existing WtE plant under construction in Sao Paulo will follow these limits. In China, the government recently established limits for mercury and other heavy metals emissions from waste incineration in the country.

Because of the need of the adequate gas cleaning systems, the investment is huge and the economic feasibility is often difficult to achieve. In Brazil, a previous study of the Sao Paulo State Energy Secretariat evaluates that the electricity generation costs are as high as USD 100 per MWh generated. This figure is extremely high for the Brazilian reality, and very much above the price paid for other electricity generation sources in the country. This is also a reality in other countries: most plants in the EU became economically

---

[1] https://eur-lex.europa.eu/legal-content/EN/TXT/?uri=LEGISSUM:l28072.

[2] http://www2.ambiente.sp.gov.br/legislacao/resolucoes-sma/resolucao-sma-79-2009-2/.

Fig. 3.6 Valorsul WtE plant, Lisbon, Portugal. *(Source: S. Coelho. Personal visit (2011).)*

feasible only due to subsidies received, both nonrefundable investments as special feed-in tariffs for the electricity generated.

Valorsul—Fig. 3.6—is an interesting example.[3] The plant had 54% of its investment funded by the EU as a nonrefundable investment. This plant processes 2000 ton/day of MSW from Lisbon and five other municipalities (LHV in a range of 7400–7800 kJ/kg) and generates 37 MW. From the total investment of €174 million (2011), €94 million were funded (nonrefundable funding) by the EU. In addition to the price paid by the customers to dispose the MSW (€20–40/ton of MSW), the electricity produced is sold by a higher tariff, equal to (2011) €84 euros/MWh ("green tariff").[4]

Regarding social concerns in the DCs, incineration plants (as other WtE plants) are often objecting of concerns regarding the lack of jobs for unskilled people. In many countries, these people usually work on existing dumps, aiming to recovery the recyclable waste to commercialize. Despite the fact that it is a hard work with several health impacts, it is seen as the only option in some poor regions. Therefore, any WtE process is seen as a threat for these workers.

---

[3] http://www.valorsul.pt/.
[4] S. Coelho. Personal communication based on the field visit to Valosul.

This is also a consequence of the lack of adequate information. WtE in the developed countries has grown simultaneously with the recycling process (*Eurostat, 2017* Avfall Sverige[5]). In the DCs, since recycling is not yet a widespread process, these unskilled workers may work in cooperatives for the recycling process, before the WtE process. This already occurs in Brazil, as discussed in several documents[6,7] and recycling cooperatives are mandatory in any WtE process under licensing in the country.

3.2.1.3 Preliminary Comments on Incineration Process

The incineration process has been used as a solid waste thermal treatment process since the beginning of the last century. During the last decades, this process has been improved, establishing environmentally reliable technologies with operations on a commercial scale. Several modern incineration plants are being built around the world including energy conversion (Menezes et al., 2000).

Among the advantages of MSW incineration, it can be highlighted the reduction of volume for disposal in landfills, the elimination of methane emissions in landfills, and the energy recovered from the combustion that can be used for the electricity production. This is an aspect particularly important in countries where energy access is still a problem.

On the other hand, waste incineration may have negative environmental impacts if not adequately controlled. There are atmospheric emissions from the process, such as heavy metals, dioxins, and furans that must be eliminated through gas cleaning systems. Therefore, countries have adopted restrict emission limits. The consequent development of advanced technologies allowed the adequate gas treatment, in order to guarantee that pollutant emissions to be under the limits. This is the main reason why this WtE process has high investment costs (IEA, 2007).

In brief, the main advantages of using incinerators are:
- direct use of thermal energy to generate steam and/or electric energy;
- continuous process of waste treatment;
- process with low noise and no odor characteristics; and
- requires small area for installation.

Among the disadvantages, we have:
- no economic feasibility for with residues of lower heating value and with halogenated ones, as well as for plants smaller than 500 ton/day of MSW;
- excessive moisture impairs combustion;
- need to use auxiliary fuels to maintain combustion in some cases;
- toxic metals can remain concentrated in the ashes;

---

[5] Swedish Waste Management, http://www.avfallsverige.se/fileadmin/uploads/forbranning_eng.pdf.

[6] http://repositorio.ipea.gov.br/bitstream/11058/7819/1/bmt_62_catadores.pdf (in Portuguese).

[7] http://www.mma.gov.br/cidades-sustentaveis/residuos-solidos/catadores-de-materiais-reciclaveis.

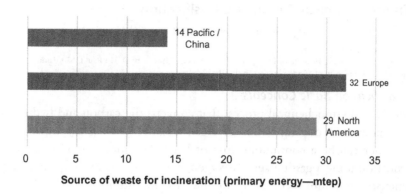

Source of waste for incineration (primary energy—mtep)

Fig. 3.7 Scenarios for MSW incineration and electricity generation in 2025. *(Data from International Energy Agency (IEA), 2010. World Energy Outlook 2010. Available from: http://www.iea.org/Textbase/npsum/weo2010sum.pdfS.)*

- high initial investment, operation, and maintenance costs; and
- toxic emissions that must be reduced to follow environmental legislation.

It is estimated that 75% of MSW generated in the OECD countries (after allowing for recycling) are available for use as fuels. This corresponds to 75 Mtoe,[8] distributed per region as shown in Fig. 3.7. Assuming a typical generation efficiency of 25%, this is equivalent to 218 TWh/year.

Prospecting studies conducted by the International Energy Agency (IEA, 2010), show the evolution of MSW availability for use in incinerators for energy conversion. Fig. 3.7 shows that, for the 2025, North America, Europe, and Asia/Oceania will account for about 29, 32, and 14 Mtoe of primary energy, respectively.

The current available technology for the incinerator design estimates the generation of up to 0.95 MWh/ton of MSW processed, with most of installed systems generating from 0.45 to 0.70 MWh/ton capacity. This production will depend on the MSW heating value (Menezes et al., 2000) in each case. These figures are confirmed by other studies. Tolmasquim (2003) estimates that it is possible to supply a 16 MW–WtE plant with 500 tons of MSW per day, which represents a specific energy generation of about 0.70 MWh/ton.

These values for sure include uncertainties about incineration estimates, including: (i) regional variation in waste composition and quantity and (ii) eventual changes in the MSW regulation and disposal (minimization in generation, reuse, and recycling), which will certainly change waste composition. However, for sure this is an interesting WtE option for large plants (above 500,000 inhabitants, in average).

---

[8] Toe = tons of oil equivalent.

3.2.2 Thermochemical Treatment: Gasification

Luciano Infiesta*, Javier Farago Escobar[†], and José R. Simões-Moreira[‡]

*Carbogas Industries, São Paulo, Brazil
[†]Research Group on Bioenergy, Institute of Energy and Environment, University of São Paulo, São Paulo, Brazil
[‡]Polytechnic School, University of São Paulo, São Paulo, Brazil

3.2.2.1 Gasification Basic Concepts

Gasification is a thermochemical process that converts the carbon and hydrogen in the chemical structures by decomposition of organic matter into a gas (syngas).

The gasifier can be a continuous flow or batch, the most common technique for partial oxidation using a gassing agent (oxygen, air, or hot steam) in quantities less than stoichiometric (minimum theoretical amount for combustion).

The production of generator gas (producer synthesis gas) called gasification is a partial combustion of solid fuel (biomass) and takes place at temperatures of about 1000°C. The reactor is called a gasifier.

Biomass gasification means incomplete combustion of biomass resulting in the production of combustible gases compound of carbon monoxide (CO), hydrogen (H_2), and traces of methane (CH_4) as shown in Table 3.10. This mixture is called *producer* gas or *synthesis* gas (syngas). Producer gas can run ICEs (both compression and spark ignition), can be used as a substitute or in combination with furnace oil for direct heat applications, and can be used to produce, in an economically viable way, methanol—an extremely attractive chemical which is useful as fuel for thermal engines as well as a chemical feedstock for industries (Reed et al., 1982).

Four distinct processes take place in a gasifier as the fuel makes its way to gasification. They are:

- drying of fuel,
- pyrolysis—a process in which tar and other volatiles are driven off,
- combustion, and
- reduction.

Table 3.10 Gasifiers' Characteristics

Characteristics	Properties
Syngas low heating value	Low: Up to 5 MJ/Nm^3 (997 kcal/kg)
	Medium: 5–10 MJ/Nm^3 (997–1993 kcal/kg)
	High: 10–40 MJ/Nm^3 (1993–7972 kcal/kg)
Type of oxidant (gasifying agent)	Air, oxygen
	Water vapor
Type of bed	Fluidized bed (bubbling or circulating)
	Fixed bed (parallel current or countercurrent)
Operation pressure	Atmospheric or pressurized (up to 6 MPa/59.2 atm)
Type of biomass	Agricultural waste, industrial or municipal (MSW)
	In natura biomass, pelletized or pulverized

Data from Fundação Estadual do Meio Ambiente (FEAM), 2010. Estudo do estado da arte e análise de viabilidade técnica, econômica e ambiental da implantação de uma usina de tratamento térmico de resíduos sólidos urbanos com geração de energia elétrica no estado de Minas Gerais: Relatório 1. 2. ed. Belo Horizonte – Minas Gerais.

Although there is a considerable overlap of the processes, each can be assumed to occupy a separate zone where fundamentally different chemical and thermal reactions take place.

The composition of the gas, with the concomitant production of solid fuels (charcoal), and condensable liquids (pyro ligneous), depends on the following factors (FEAM, 2010):

— type of gasification reactor,
— power supply to the process,
— introduction or not of water vapor with the oxidant (air or O_2),
— retention time, and
— gas withdrawal system and other products used in organic matter.

Table 3.10 illustrates main gasifiers' characteristics, based on FEAM (2010).

In most currently installed systems, the gasifier agent is air, producing a syngas with low-medium heating value. In addition, all systems in operation are at atmospheric pressure, as the large pressurized systems have been disabled.

The requirements of the biomass fueled depend on the type of gasifier. Fixed bed gasifiers require a suitable particle size and humidity of 25%. For less dense biomass, these gasifiers require it to be pelletized, which in most cases makes a difficult energy and economic balance. Fluidized-bed gasifiers accept biomass with lower density and higher humidity.

Current situation of existing gasification plants worldwide is presented at Task 33 of IEA Bioenergy.[9] In this site, we can note that nowadays there are no plants using gas turbines for electricity production, since all existing plants with this technology were shutdown due to economic and technical problem (Hrbek, 2015).

Regarding MSW gasification, there are only a few plants in industrialized countries; in the DCs, there are pilot plants in Brazil, as presented ahead in this section. Table 3.11 shows the average composition of syngas.

Syngas can be used in several applications, such as electricity generation, heat production for direct heating, and as a raw material for obtaining liquid fuels through the Fischer–Tropsch synthesis (methanol ethanol, ammonia, gasoline, diesel, etc.).

Table 3.11 Average Composition of Syngas

Components	Concentration (%)
CO	8 a 25
H_2	13 a 15
CH_4	3 a 9
CO_2	5 a 10
N2	45 a 54
H_2O	10 a 15

Data from Reed, T.B., Graboski, M., Markson, M., 1982. The SERI High Pressure Oxygen Gasifier. Report SERI/TP-234-1-455R, Feb., Solar Energy Research Institute, Golden, Colorado.

[9] http://task33.ieabioenergy.com/.

3.2.2.2 Main Types of Commercialized Gasifiers

The main types of gasifiers currently commercialized are the *fixed bed* and the *fluidized-bed* gasifiers. The choice of the type of gasifier depends upon the biomass characteristics and the size of the unit.

Fixed Bed Gasifiers

Indian companies are those who mostly commercialize small systems with fixed bed technology. The units with power from 1 to 200 kW operate in most cases on diesel fueling dual system (syngas-diesel[10]), but there are some experiments with a gas engine adapted to syngas. These systems are easy to operate and can be used for small-scale power generation in rural areas and isolated systems in the DCs. There are several examples of systems installed in India, China, Cuba, and Brazil (IISc, 2010), Brazil/Amazon (CENBIO/IEE/USP, 2006), and in Cuba (GEF/UNIDO, 2014).[11]

The material to be gasified moves by gravity. Such gasifiers are built so the solid biomass forms the fixed bed, supported by a grid. Fixed bed gasifiers can be upflow (updraft) or down (downdraft) with respect to the syngas and air feeding. This is the most widespread technology, quite well known and dominated operationally, being implemented mainly at small scales. However, in the case of MSW, it must be dried and pelletized before being fed into the gasifier, which increases overall energy consumption and costs.

Fixed bed gasifiers are used with ICEs, in capacities from 1 up to 200 kW, considering the adequate operation. Fig. 3.8 illustrates fixed bed gasifiers.

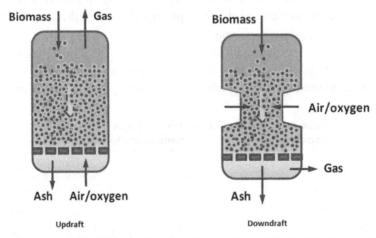

Fig. 3.8 Updraft and downdraft fixed bed gasifier. *(Adapted from E4Tech, 2009. Review of Technologies for Gasification of Biomass and Wastes. Final report. NNFCC Project 09/008, June.)*

---

[10] 80% syngas, 20% diesel.
[11] Project GEF/UNIDO. Unit with 50 kW in Cocodrilo, Isla de la Juventud, (in operation) (visit by S. Coelho, 2014).

In downdraft gasifiers, the biomass is fed at the top of the gasifier and the air, or oxygen and/or steam intake is also at the top or from the sides; hence the biomass and gases move in the same direction. Some of the biomass is burnt, falling through the gasifier throat to form a bed of hot charcoal, where the gases have to pass through (the reaction zone). This ensures a higher quality syngas with low tar and particulates, which leaves at the base of the gasifier, with ash collected under the grate. The low yield (around 15%–20%), the difficulty of handling (manual feed), and the ashes generated in the process are common problems in these small downdraft gasifiers (CENBIO/IEE/ USP, 2006).

In the updraft fixed bed gasifiers, the biomass is fed in at the top of the reactor, and the air, oxygen, and/or steam intake is at the bottom, hence the biomass and gases move in the opposite directions. Some of the resulting char falls and burns to provide heat, and the methane and tar-rich gas leaves at the top of the gasifier. Air and steam are injected to keep the ash below the melting temperature and facilitate coal conversion. The ash falls from the grate for collection at the bottom of the gasifier, producing a syngas with little PM but with higher tar content (10%–20%). For applications in ICEs, this tar must be removed.

Fluidized-Bed Gasifiers

This type of equipment uses a material kept suspended in a bed of inert particles (sand, ash, or alumina) and fluidized by the air stream, which drags the biomass. It can be the bubbling or circulating type as the speed at which the material passes through the bed.

Fluidized bed is mainly used for power production higher than 200 kW, but it can also be used for smaller scales. These reactors allow the use of lower density and higher humidity biomass.

In the 1990s, there were several initiatives in Europe and the United States, developing pilot plants aiming to produce the syngas to feed gas turbines mainly in combined cycles (the so-called IGCC, integrated gasifier combined cycle). Pilot plants in Varnamo (Sweden), Maui (Hawai), and in the United Kingdom (ARBRE project), among others, expected to develop both pressurized and atmospheric gasifiers to be integrated to gas turbines. However, technical problems—mainly due to the difficult to adequately syngas cleaning to be fed into the gas turbine—and economic difficulties obliged all plants to be shutdown. Nowadays, as shown in Task 33 from IEA Bioenergy, the existing plants in operation are mainly for thermal purposes or for power production in engines, where there is no need for a high-quality syngas.

Fluidized-bed gasifiers can be classified as pressurized or atmospheric gasifiers (depending on the working pressure) or as bubbling or circulating bed gasifiers (depending on the velocity of the fluidized bed).

In the *bubbling fluidized bed (BFB),* the bed is made of fine inert material. At the gasifier bottom, air, oxygen, or steam is blown upwards through the bed just fast enough (1–3 m/s)

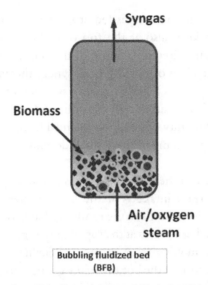

Fig. 3.9 Bubbling fluidized-bed gasifier. *(Adapted from E4Tech (2009).)*

to agitate the material. Biomass is fed in from the side, mixes, and forms syngas, which leave upwards. The system operates at temperatures below 900°C to avoid ash melting and sticking. The system can be pressurized or atmospheric. Fig. 3.9 illustrates a BFB gasifier.

In the *circulating fluidized bed (CFB)*, there is the same bed of fine inert material with air, oxygen, or steam blown upwards, but it is fed in a higher velocity (5–10 m/s), to suspend material throughout the gasifier, allowing a better mixing of fuel and air. Biomass is fed in from the side and reacts to form the syngas. The mixture of syngas and particulate matter are separated using a cyclone, with material returned into the base of the gasifier. It operates also at temperatures below 900°C to avoid ash melting and sticking and can be pressurized. Fig. 3.10 illustrates a CFB gasifier.

These types of gasifiers are more adequate to the conversion of large amounts of biomass; systems with a capacity of 10–20 tons of biomass per hour are already operational. They are also more flexible as to the feedstock characteristics and can be used to convert biomass with minimal processing requirements.

Gasification systems present great advantages over small plants for power generation, especially in the rural areas and isolated regions, as well as for the disposal of MSW in small towns. In addition, there are other benefits:

- Gasification has a higher energy efficiency (65%–80%) compared to direct combustion (60%–75%).
- Ashes and residual carbon remain in the gasifier, thus reducing particulate emissions.
- It is easily distributed.
- Syngas burning is more easily controlled.

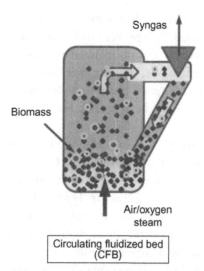

Fig. 3.10 Circulating fluidized-bed gasifier. *(Adapted from E4Tech, 2009. Review of Technologies for Gasification of Biomass and Wastes. Final report. NNFCC Project 09/008, June.)*

3.2.2.3 MSW Gasification

WtE through gasification appears to be a very attractive option, due to two simultaneous processes: electricity production and the adequate final MSW disposal, in an environmentally appropriate process.

The process of MSW gasification is divided into three stages:
- the transformation of the MSW into a more homogeneous fuel (RDF);
- the gasification process, occurring inside the thermochemical reactor; and
- use of the heating value from the synthesis gas for energy conversion.

At the first stage, the MSW is received and sent to a sequence of automatic operations, with the objective of separating undesirable and recycling materials, like metals (ferrous and nonferrous), glasses, stones, and inert components. This process also reduces the waste humidity down to the adequate levels, ensuring a final fuel with both constant granulometry and adequate heating value, called refuse derived fuel (RDF). The next stage corresponds to the main part of the process, the gasification, inside the thermochemical reactor. This process converts RDF into the syngas, a combustible gas with multiple applications. During this gasification phase, the RDF passes through a chemical cracking process, producing the syngas, composed of CO, CO_2, H_2, and CH_4.

As mentioned above, the residues from the gasification process include nitrogen and small particles of ashes. The MSW syngas is a gas with quite reduced heating value and, for a typical system, it is in the range of 1100–1200 kcal/Nm^3. For MSW/RDF, the fluidized-bed gasifier appears to be the most adequate, since it favors the mass and heat transference (INFIESTA, 2015).

At the final step, it is located the electricity generation, which is one of the possible applications for the synthesis gas. The power production can occur in two different technological options nowadays commercialized: in an otto-engine adapted for syngas or in a Rankine steam cycle. In the first one, syngas can be used as a fuel directly in the engine; otherwise, in the steam cycle, syngas is burnt in boilers and the steam is fed into turbogenerators. The whole plant can run in an automatic operation, leading to great efficiency and security.

In São Paulo, Brazil, Carbogas Co. has developed a RDF-fluidized-bed gasifier and operates a 1 MWth pilot plant, as shown in Fig. 3.11. In this plant, there are specially projected cyclones inside the gasifier to guarantee that nonconverted carbonaceous will react. Therefore, there will be no residual tar. Hydraulic valves allow the fuel feeding without any fugitive gas emissions to the environment (INFIESTA, 2015).

The gasification reactor must operate in an incomplete combustion process (to allow the production of the syngas), in a sub-stoichiometric scenario, with oxygen concentrations below the minimum theoretical for combustion. This is why there is no synthesis of pollutants like dioxins and furans, since their formation requires oxygen; their formation is not possible due to a chemically unavailability, leading to a clean and effective process of power generation and solid waste disposal.

3.2.2.4 Syngas Energy Conversion Technologies

Syngas composition is typically dominated by the following combustible gases CO, CH_4, and H_2, altogether making up from 60% to 70% of the total composition, along with 30%–40% of inert gases N_2 and CO_2 at different compositions depending on the feedstock. Other undesirable gases may also be present that demands further cleaning stages. Naturally, the syngas heating power is lower than the ones of plain carbon monoxide,

Fig. 3.11 MSW gasification pilot plant (Carbogas, Mauá/SP, Brazil) and MSW gasification industrial plant in Furnas, Boa Esperança/MG, Brazil. *(Source: Photos taken by the author (2018).)*

methane, or hydrogen alone due to the presence of those inert gases. Nevertheless, it is a good fuel and it can be either directly burned in a boiler to produce steam to drive a Rankine cycle or as a fuel for powering ICEs or gas turbines as already discussed earlier.

Base load large-scale electrical power production (tens of megawatts) is usually based on Rankine cycles. In those cycles, the overall efficiency is in the range 30%–35%. Typically, burning syngas in the boiler generates steam to feed a steam turbine that drives an electrical generator. Such plant configuration allows a continuous electrical power generation following the syngas production and can have an electrical power output up to 300 MW. The main cycle components are shown in Fig. 3.12A. Depending on the plant

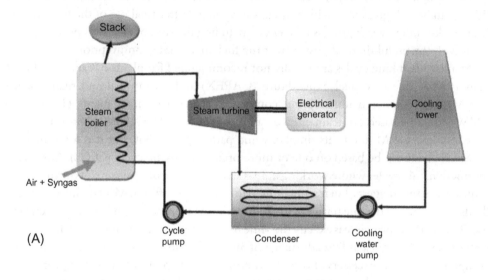

(A)

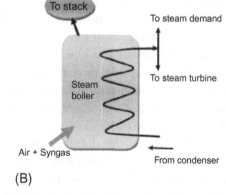

(B)

Fig. 3.12 Rankine cycle power plant (A) standard Rankine cycle plant and (B) cogeneration or CHP configuration.

necessities, the system can also operate in a cogeneration or (CHP) configuration so that steam can also be produced to be used for other demands besides electrical power generation as illustrated in Fig. 3.12B. Regarding that figure, in many configurations, steam is derived from a medium to a low turbine stage, at a lower pressure, to supply to steam demand instead of high-pressure derivation as illustrated in Fig. 3.12B. A perfectly balanced cogeneration system will provide steam to both electrical power generation and for heating or any other process. Usually, at night the electrical power demand drops and thus operational adjustments are necessary if steam is still demanded for a full-time operating plant, considering that syngas production is at a 24 h per day rate. In cogeneration systems, a higher use of the syngas energy content is obtained as the energy utilization factor (EUF) can be as high as 80% which implies in a much better final use of the fuel energy. EUF is a kind the overall process efficiency, including the sum of electrical power generation with the useful thermal power over the fuel energy rate consumption.

Standard Rankine cycles are usually not recommended for plants of capacities lower than 10 MW due to capital expenditure (CAPEX) and operational and maintenance (O&M) costs. Some manufacturing companies claim that small plants (100 kW to 1 MW) could be based on steam engines, an old and low CAPEX device. Steam engines have also low O&M due to its simplicity and parts replacement, but it is less efficient.

Small plants can be based on one or more organic Rankine cycle (ORC). ORCs are thermodynamic cycles whose working fluid is an organic one or a mixture of organic fluids as the cycle name suggests, but it is not limited to only organic fluids. Most of those working fluids are widely used in the refrigeration industry; however, other fluids are also in current use. The working principle is exactly the same to the conventional Rankine cycle based on water steam. The outstanding advantage of an ORC is its simpler operation and many components are not necessary. From the thermodynamic point of view, the problem of blade erosion caused by wet vapor expansion (water droplets) is either eliminated or reduced because of some special working fluid properties. Thus, such fluids do not need a superheating stage in the boiler as saturated vapor can directly feed the turbine. Typically, ORCs are used as a way to produce additional electrical power by heat recovery from industrial thermal processes or stack gases. It can be used in remote systems and it is more flexible to variable load than the standard Rankine cycle. Capacity ranges from tens of kilowatts to a few megawatts. Some companies now deliver off-the-shelf ORCs. Overall efficiency is usually low due to the low operational temperature range.

Syngas can also fuel stationary ICEs for electricity production. Arrangements of one or a set of ICEs can be engineered in order to adjust the electrical demand so that a broad range of power can be supplied, from a few hundred kilowatts to 20–30 MW, which give them some flexibility. The basic configuration is shown in Fig. 3.13A. When compared to gas turbines, it has the additional advantage of requiring a low or near-atmospheric syngas feeding pressure. ICEs also are more flexible to syngas composition and contamination. Also, as pointed out by some manufacturers, hydrogen gas combustion is

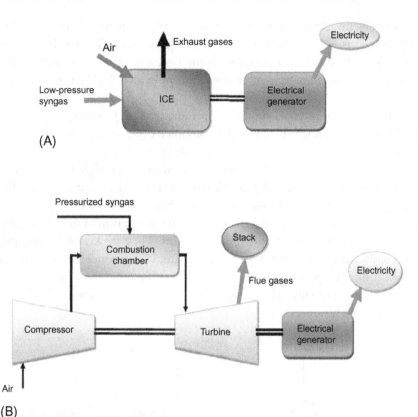

Fig. 3.13 (A) Internal combustion engine (ICE) and (B) gas turbine.

considerably faster than carbon monoxide, which requires special engine adjustments to avoid operational problems such as knocking.

ICEs can be categorized in two main types: spark ignition and compression ignition. In spark ignition, fuel and air are mixed prior to been aspired to the combustion chamber. Light oils and diesel are the fuels used in compression ignition. In this engine, only air is compressed in the combustion chamber to a crank angle where the fuel is injected in the chamber. The first technology can work directly with syngas or in combination with natural gas, for instance. The later technology is usually operated in dual fuel mode, where diesel (or light oil) is one of the fuels and the syngas is mixed with admitting air. ICE usually has lower CAPEX, low O&M and low cost parts replacement, and have a high modularity.

Gas turbine is the last conventional technology for electricity power production from syngas as illustrated in Fig. 3.13B. Base-load large plants are usually engineered because of the large investment (CAPEX and O&M) demanded. When compared to natural gas fired gas turbines some technological adjustments are necessary because of the large hydrogen content in syngas composition. As stated above, hydrogen fast burning does pose some problems. Besides that, a higher fuel mass flux is required of syngas when compared to natural gas, because the presence of inert gases. Manufacturers have solved these technological problems. Other backdrop is the necessity of syngas compression, which demands nonnegligible amount of compression power. ICEs do not need such a gas compression stage.

Flue gases leave gas turbines at high temperature, which brings about a configuration well used in power production, that is, the combined cycle. A combined cycle has one or more gas turbines producing power along with steam-based cycle (Rankine). A heat recovery steam generator (HRSG) produces steam from the high-temperature exhaust gas to power a Rankine cycle. High efficiency is reached in combined configuration (50%–60%). Fig. 3.14 shows an illustration of a simple combined cycle.

Finally, large plants in operation are based on the integration of the whole gasification process along with the electrical power production. They are known as integrated gasification combined cycle (IGCC). In the same site, it is produced syngas and electrical power by a combined gas turbine—Rankine cycle. Of course, there are advantages from the operational point of view. However, the problem of syngas cleaning has been a major hurdle for large plants operation.

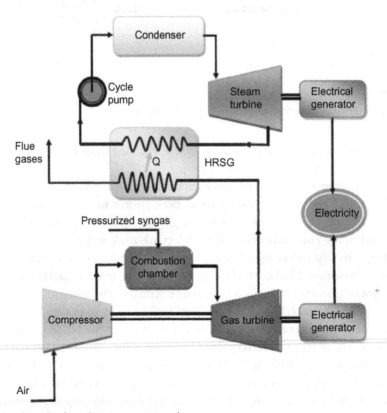

Fig. 3.14 Combine Rankine-Brayton power cycles.

3.3 CURRENT SITUATION OF WtE IN LATIN AMERICA
3.3.1 Current Situation of WtE in Argentina

Estela Santalla

Departamento Ingeniería Química, Facultad de Ingeniería/UNICEN, Buenos Aires, Argentina

The current scenario of WtE projects in Argentina describes the sectors and biomass/residues involved, technological characteristics and policy framework. The historical contribution of biogas and LFG in the electricity market of Argentina is also analyzed. The impact of clean development mechanisms (CDM) and the policy framework of promotion of renewable energy for electricity were evaluated also considering the crucial environmental problems related to the inadequate management of agro-industrial residues added to an outlook of energy restrictions, which reveal a promissory scenario for the development of the bioenergy in the country.

Biogas can be generated in landfills and in digesters. LFG is produced from the organic waste degradation in MSW final disposal sites through a less controlled anaerobic process influenced by climate conditions (environmental temperature and rainfall), and by the operative conditions of the landfill such as compaction, waste coverage, and leachate treatment, between others. Biogas produced in digesters is the results of a biochemical-controlled process where technical (such as the volumetric organic load, residence time, and the reactor design) and management (availability of residues/biomass, logistic, etc.) issues play key roles on the optimal performance of the methane production.

The analysis of the state of art of the technology in Argentina reveals that biogas implementation showed slower growth than LFG, due to the influence of CDM scenario, since from 2004, this mechanism of the Kyoto Protocol allowed to the DCs to achieve GHG certifications through the methane capture in landfills. From 1980, the high concentration of population in the Metropolitan Area of Buenos Aires induced to concentrate the MSW in large disposal sites that gradually become in landfills. Therefore, when the Kyoto Protocol went into effect, 11 LFG projects between small and large scale[12] were formulated and registered under this mechanism with the main purpose to GHG mitigation. From the seven projects that achieved the issuance of certified emission reductions (CERs), the performance in terms of the ratio between expected and actual methane destroyed varied between 4.3% and 53.4%. The average (28%) coincided with several studies that explained problems related to overestimation of the LFG emission models, monitoring of methane fraction and management, and operation of MSW disposal sites, between others (Blanco et al., 2016).

The development of biodigestion, on the other hand, went through a set of technical and institutional barriers not yet overcome (see Section 3.2.2). Under the CDM mechanism, five projects achieved the registration of GHG mitigation by the methane capture

---

[12] https://cdm.unfccc.int/Projects/projsearch.html.

from anaerobic wastewater treatment in the agro-industrial sector (starch, yeast and fruit processing and poultry slaughterhouse), using covered lagoons. Although none could achieve issuance of CERs. Although, thanks to the institutional framework of renewable energy promotion, several projects of biogas for electricity based on the use of agro-industrial residues and biomass such as swine wastewater, corn, and vinasse silages are progressively entering into operation.

Regarding technological transfer issues, the contribution of CDM on LFG projects resulted limited, since technological learning and capacity building were only limited at the level of O&M of foreign technologies, situation evidenced in 45% of the whole of 44 CDM registered projects (Blanco et al., 2016). Regarding biodigestion projects, the results of a national survey revealed that prevails local construction companies (with foreign equipment) and in some cases, the implemented technological models did not adapt to local conditions, having plants that run poorly and/or with higher operating costs (INTI, 2016).

A first approach of some sectors considers the biodigestion as a simple technique that always will produce biogas. They leave aside the understanding that a biologic process requires specific and controlled conditions, added to insufficient technological and human resources for a real technologic development and an unequal competition with other renewable energies such as wind and photovoltaic are obstacles that delayed the growth of this technology in Argentina.

The official PROBIOMASA Program[13] describes a scenario where there are problems that could be switched in an opportunity for developing the biodigestion technology:

  (i) the increase of the productivity of the agricultural systems without the consequent increase of wastewater treatment,
 (ii) the lack of sustainable consciousness,
(iii) the absence of penalties,
 (iv) a weak structure of economic incentives for the implementation of sustainable systems, and
  (v) difficulties in the certification of the coproducts and to access to information.

A previous work developed under the Technological Needs Assessment[14] identified similar barriers as described above, in addition to the higher costs of the technology of biodigestion against wind and solar ones and the scarce number of local suppliers that improve competitiveness issues (TNA, 2012).

A recent report revealed that most of the small and medium plants showed several technical shortcomings in terms of construction, materials used, safety, and operation; there are also many facilities that operate in an elementary way (they make decisions based on the visual aspects and experience) that are willing for advice and technical assistance.

---

[13] http://www.probiomasa.gob.ar/_pdf/Capacitacion%20%202016.pdf.
[14] Implemented by the United Nations Environment Programme and the UNEP DTU Partnership on behalf of the Global Environment Facility http://www.tech-action.org/.

With few exceptions, anaerobic biodigestion plants do not present safety measures appropriate to the type of process being a critical point of improvement.

The use of by-products (as organic fertilizer) is not yet developed in the country due to multiple factors, mainly the lack of clarity in the regulations for its application, which directly impacts the bottom-up flow of projects in decreasing their feasibility since it gives a potential development of sustainable fertilizers to agriculture, decreasing the use of mineral ones.

On the part of the users, there are shortcomings of specific knowledge about biogas technology and there is a great lack of knowledge in the technical and professional management of the facilities of biogas plants. Unsuccessful cases were basically based on the lack of knowledge from users, on technologies that were not applicable to the climatic and geographical conditions of plant location and technological failures mainly in materials used and process conditions (INTI, 2016).

3.3.2 Current Situation of WtE in Brazil

Suani Teixeira Coelho

Research Group on Bioenergy, Institute of Energy and Environment, University of São Paulo, São Paulo, Brazil

In Brazil, electricity is produced from biogas conversion in landfills using engines. There are also some WtE local initiatives with both MSW incineration (in São Paulo, the incineration plant at Barueri municipality) and MSW/RDF gasification in São Paulo and in Minas Gerais states, as discussed ahead. Incineration technology is mainly imported since there is no local manufacturer and presents high costs. Gasification technology under implementation is locally developed but yet being further developed. Existing experiences show gasification technology is the better option for small municipalities since waste incineration is not commercially feasible for power production below 5–10 MWe.

In general, MSW gasification is expected to be cheaper than incineration because there is no needed for high-level gas cleaning system. Gasification operates with low oxygen to allow the incomplete combustion (gasification) process and because of that, dioxins and furans production is extremely low, with no need for additional (and expensive) cleaning systems.

However, these are few initiatives and further policies are needed since the main difficulty seems to be the economic feasibility of the WtE technologies using thermal treatment processes.

3.3.3 Current Situation of WtE in Mexico

Gustavo Solórzano

Asociación Mexicana de Ingeniería, Ciencia y Gestión Ambiental, A.C. (AMICA), Mexico City, Mexico

Given the potential advantages of this kind of projects, and the volume of MSW generated every day but also partially accumulated in Mexico's almost 2500 municipalities, it seems that eight energy-producing landfill projects are a small number. However, it is expected that the recent energy reform and an improving regulatory framework will stimulate these and other WtE projects in Mexico.

3.3.3.1 Coprocessing MSW

For many years, cement kilns in Mexico have been coprocessing wastes such as spent oil/solvents, paint, tires, and, at a lesser degree, MSW. The Mexican cement industry reports coprocessing levels of 12% in terms of thermal energy replacement (GIZ-EnRes, 2016).

Among others, cement producer CEMEX started a project with Mexico City in 2012, in which it receives a mix from the inorganic fraction in MSW, with an average composition of 32% plastics; 50% paper and cardboard; 10% textiles; and 8% wood. CEMEX prepares and transports the waste to two cement plants and feeds the waste into its own cement kilns. A total of 84,000 tons of inorganic MSW were processed in eight CEMEX plants in the country in 2013 (GIZ-EnRes, n.d.).

According to the German International Agency (GIZ), there is a potential replacement for thermal energy in the cement industry up to 30% in Mexico, which means using 3.1 million tons per year of MSW. This would be equivalent to 8.1% of the total amount generated in the country, which has the potential to mitigate the emission of 3.2 Mt of CO_2e per year. Nevertheless, a lack of source separation programs for MSW, pretreatment/transportation costs, and an inadequate regulatory framework are important challenges to this WtE alternative (GIZ-EnRes, 2016).

Anaerobic Digestion

A significant number of anaerobic digestors (currently around 700) have been operating in Mexico with different kinds of animal excreta for several years. But the digestion of the organic fraction in MSW (OFMSW) as an option to treating this fraction in MSW has only recently been explored, as is the case in other countries in LAC (Latin America and Caribbean) region.

The first anaerobic digester for OFMSW has been built in Atlacomulco, Mexico, with a capacity to process 30 ton/day. The digester was built with a US$2.8 million grant from the federal government; land was provided by the municipality. The digestor contemplates a wet process and according to the designer it will produce up to $2400\,m^3$/day of methane with an energy yield of up to 1,752,091 kWh/year (GIZ-EnRes, n.d.).

Incineration

Mexico City (CDMX) has initiated the process of building and operating an incineration WtE plant. In December 2016, CDMX published the "specifications for the award of the contract for … the design, construction, commissioning, operation and maintenance of a plant for recovery of the calorific value of municipal solid waste … for the generation and delivery … of electricity … for up to 965,000 MWH a year" (CDMX, 2017). The bidding document was expressly written this way due to the city government's aim to stop paying for the energy needed to run its underground transportation system, transporting almost 6 million passengers every day. The corresponding contract, valid for 33 years, has been granted to the multinational French firm Veolia, who will operate the plant to burn nearly 4500 ton/day out of the 13,000 tons of MSW generated daily, of which 8500 ton are currently being exported to landfills located outside the city (Veolia, 2017). As such, the plant is designed to provide the energy requirements to run the underground transportation system, an action that according to the CDMX and Veolia will mitigate the equivalent of 700,000 tons of CO_2 emissions. The original site where the plant would be built, next to the new international Mexico City airport (now under construction) has been reconsidered and CDMX is currently defining a new site near the originally proposed one.

3.4 CURRENT SITUATION OF WtE IN ASIA

3.4.1 Waste Incineration in Asia

Roshni Mary Sebastian, Dinesh Kumar, and Babu J. Alappat
Indian Institute of Technology, New Delhi, India

One of the most crucial elements of waste incineration is the input, that is, MSW feed. Deciding if the feed must be incinerated on "as-received" basis (mass burning) or after pretreatment (drying/shredding/sorting/refuse derived fuel (RDF) production) plays a major role in the choice of the incineration technology.

Mass burning with moving grate incinerator is most commonly used in Asia, due to its capability to accommodate MSW feed with fluctuating composition and heat content. Fluidized-bed technology finds application, especially in WtE plants, which incinerate MSW with high moisture content. Chinese MSW incinerators now rely on CFB as the dioxin and furan emissions from CFBs are found to be much lower than the EU standards (World Energy Council, 2016). The incineration technology adopted should also ensure minimal pollutant emissions. Adequate flue gas retention time, mixing and recirculation of flue gases, supply of combustion air, and mixing of waste feed to ensure complete combustion are a few measures, which can minimize the pollutant emissions.

One of the significant benefits of waste incineration is energy recovery. The efficiency of an incineration plant is determined by the potential application of the recovered energy. While the production of electricity might reduce the net operational efficiency but has significant revenue generation, heat recovery for district heating is more efficient with lower revenue generation. The flue gases released at 1000–1200°C are cooled by boiler systems. The energy recovery system may be used to produce hot water, low-pressure steam, or electricity. Hot water with temperature in the range of 110–160°C can be generated while the temperature of the steam generated may be about 120–250°C. Fig. 3.15 shows the layout and details of a WtE plant.

The most critical element of concern in opting incineration for MSW treatment and disposal is the pollution potential of the process. The composition of the waste feed and the combustion technology determine the pollutant release. Particulate, SO_2, NO_X,

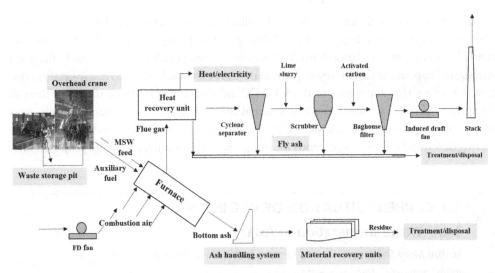

Fig. 3.15 A closer view of the various units of aWtE plant in Asia. (Data from Hitachi Zosen, 2014. http://www.mofa.go.jp/region/latin/fealac/pdfs/4-9_jase.pdf. Accessed 19 April 2018.)

HCl, HF, dioxins, furans, and heavy metal emissions may be present, which requires advanced air pollution control equipment. Besides primary control measures entailing modification of the combustion technology, secondary control measures like use of electrostatic precipitators, baghouse filters, etc. may be used. The air pollution control measures must be designed to adhere to the respective emission standards (Rand et al., 2000).

3.4.2 Characteristics of WtE in Asia

Roshni Mary Sebastian, Dinesh Kumar, and Babu J. Alappat
Indian Institute of Technology, New Delhi, India

The composition and characteristics of the MSW, influenced by the rate of urbanization, lifestyle habits, consumption patterns, cultural differences, and local climatic conditions, need to be conducive for incineration to be a feasible treatment option. Incineration may be classified as a disposal technique or an energy recovery technique based on the energy efficiency of the incinerator (EC, 2008). The incinerability of the MSW feed may be crucial to the feasibility of incineration, irrespective of the objective of the process. Incinerability is the amenability of MSW to be burned completely to sterile ash, with minimal environmental impact, optimum energy recovery, and economic sustainability (Sebastian et al., 2017).

While quantification of incinerability may remain indeterminate, some conventional methods have been adopted to ascertain the feasibility of incineration (Sebastian and Alappat, 2016). Tanner diagram (see Fig. 3.16) is one such method used for the analysis

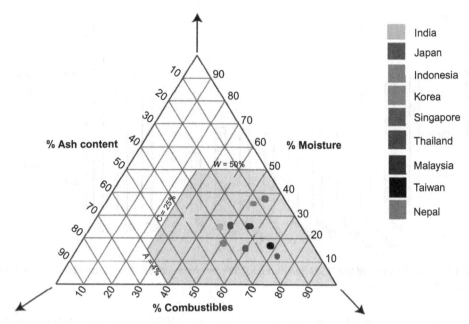

Fig. 3.16 Position of the MSW generated in different countries in Tanner diagram.

of the self-sustained combustibility of MSW. The position within the Tanner diagram, which is determined by the proximate analysis of the MSW feed, is used to determine the self-sustained combustibility of MSW (Tanner, 1965).

The calorific value of the MSW has also been used for deciding the viability of the technique. A calorific value of <1500 kcal/kg roughly is deemed to be nonincinerable, while more than 2400 kcal/kg approximately is said to be autogenously incinerable. Auxiliary fuel may be used to sustain the combustion in the intermediary range (Rand et al., 2000). Since the incinerability of MSW need to account for the environmental impact of the process as well, which is neglected in the conventional methods, there is a pronounced need for a decision-making tool, which can quantify the incinerability of MSW incorporating the 3-E concept. Studies are being carried out to bridge this gap. Fig. 3.17 illustrates the self-sustained combustibility assessment using conventional techniques.

3.4.2.1 Incinerability of MSW Generated in South East Asia
The incinerability of MSW generated in various Asian countries was estimated using the recently formulated incinerability index or *i*-index on a scale of 0–100 (Sebastian et al., 2019a,b,c), as shown in Fig. 3.18. As opposed to the conventional methods, this technique incorporates the potential environmental impact, energy recovery as well as economic aspects to compute the incinerability of MSW. Using the MSW composition data,

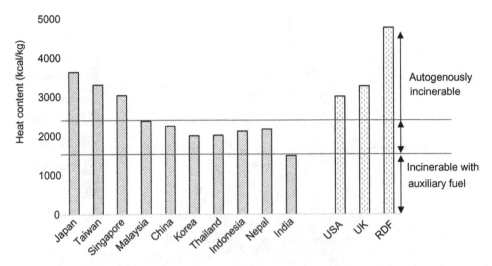

Fig. 3.17 Calorific value of the MSW generated in different Asian countries and their comparison with the developed countries.

the incinerability was quantified theoretically and is illustrated in Fig. 3.18. It is evident that high-income countries like Japan, Singapore, and Taiwan have relatively high incinerability, apparent from the indicator values of 77.7, 73.7, and 74.9, respectively, while MSW generated in developing economies like India, Nepal, Dhaka (Bangladesh), and Yangon (Myanmar) are significantly lower in comparison, as evident from the indicator values of 46.9, 46, 55, and 51.1, respectively. Higher fraction of wet organics and inert components and the resultant high moisture content and low calorific value reduces the incinerability of MSW considerably, which is reflected by the quantified incinerability. The comparative assessment of MSW generated in different economies has been discussed by Sebastian et al. (2018).

3.5 CURRENT SITUATION OF WtE IN AFRICA

William H.L. Stafford

Council for Scientific and Industrial Research, Stellenbosch, South Africa; Department of Industrial Engineering, University of Stellenbosch, South Africa

In Africa, approximately 0.65 kg of MSW is generated per person per day, and this is expected to reach 0.85 kg by 2025 (Hoornweg and Bhada-Tata, 2012a,b). Waste generation varies considerably between urban and rural areas, and the composition of waste varies depending on the household income and the per capita consumption levels. Generally, the percentage of organic content degreases with increasing per capita income due to a greater amount of paper, bottles packaging present in the waste streams. Africa's rapid waste generation growth rate (30% between 2012 and 2025) is largely driven by urbanization and increased wealth and is not expected to stabilize before 2100 (Hoornweg et al., 2015).

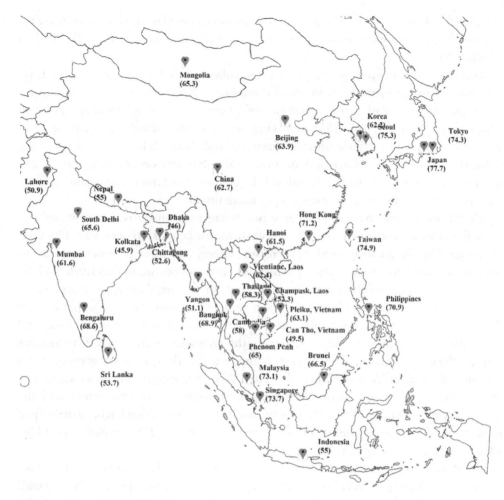

Fig. 3.18 Incinerability of MSW generated in South East Asian countries on a scale of 0–100. *(Data from Sebastian, R.M., Kumar, D., Alappat, B.J., 2019a. Identifying appropriate aggregation technique for incinerability index. Environ. Prog. Sustain. Energy 38 (3), pp. 1–10. https://doi.org/10.1002/ep.13068; Sebastian, R.M., Kumar, D., Alappat, B.J., 2019b. An easy estimation of mixed municipal solid waste characteristics from component analysis. J. Environ. Eng. https://doi.org/10.1061/(ASCE)EE.1943-7870.0001588; Sebastian, R.M., Kumar, D., Alappat, B.J., 2019c. Demonstration of estimation of incinerability of municipal solid waste using incinerability index. Environ. Dev. Sustain. https://doi.org/10.1007/s10668-019-00407-3.)*

However, the adequate collection and disposal of waste through municipal collection and treatment services remains a significant challenge in most African towns and cities. In low- and middle-income countries, solid waste management can be the single largest budget item on the municipal budget consuming 20%–50% of the annual municipal budget (Kubanza and Simatele, 2016).

In industrialized countries, most MSW and sewage are collected and disposed of appropriately in a manner that often combines energy generation with the recycling of materials. However, most African countries are developing fast and therefore, there is a lack of investment in waste management, with collection rates ranging from 25% to 70% (UNEP, 2015).

Due to the lack of appropriate public waste collection and treatment infrastructure, wastes are typically managed as informal dumpsites, often with burning as a management practice. Similarly, the provision of energy services is generally poor in Africa—with only 30% of Africans having access to municipal electricity services and 80% relying on traditional biomass (firewood and charcoal) for household cooking needs (IEA, 2011a,b). The demand for municipal public services of energy supply and waste treatment are set to increase, with the development and industrialization of Africa with the populations becomes increasingly more urban.

Despite the obvious synergy between waste management and provision of energy services, few African countries have exploited waste for the generation of energy. This is set to change with the growing need for increasing energy and waste treatment services to fulfill the demands for the increasing populations, increasing consumption levels, and the migration of Africa's rapidly developing towns and cities. Many African governments are urgently seeking integrated solutions for the treatment of wastes. These solutions include the banning of plastic bags that often litter the environment and take up valuable land space, the recovery and recycling of materials, the avoidance of emissions at landfill sites through the flaring or harnessing LFG for energy, as well as the generation of energy from organic and combustible wastes using dedicated WtE technologies. Given a long history of landfilling wastes in many African town and cities and the focus on the options with the lowest investment of capital cost, the most common opportunity that has been developed is the flaring of LFG or the capture of LFG for energy, and the CDM have often provided added financial incentives for these projects.

In several in African countries, such as South Africa, Uganda, Kenya, and Zimbabwe (discussed in Chapter 5), there are recent projects or current plans to produce energy with the treatment of both sewage and the organic fraction of MSW. Aside from increasing the supply of public services, WtE projects significantly reduce the impacts on the environment in terms of avoiding and mitigating GHG emissions and reducing the impacts from waste pollution.

In South Africa, where municipal waste collection rates are over 60%, there are several WtE projects. One example is the biogas production from the organic fraction of MSW in the city of Cape Town with the sale of approximately 600 GJ/year of compressed biomethane an industrial substitute for natural gas. In Johannesburg, there is also the use of sewage for biogas production to provide CHP for on-site wastewater treatment at municipal wastewater treatment plants. Finally, there is the capture of LFG for electricity generation in a range of 1–6 MW at several landfill sites in Durban and Johannesburg.

In many parts of Africa, incineration is not seen as a favorable option for formal waste disposal due to concerns of air pollution. However, there are many modern incineration plants in the developing world that are operating efficiently with low air emissions to meet strict air emission regulations such as in the EU. Recently, Addis Ababa, Ethiopia, has developed a WtE project, which incinerates 1400 tons of waste per day and generates 55 MW electricity (Othiang, 2018).

However, in many African countries, a large proportion of the population is still rural and wastes are managed on an individual or community basis. This has led to some local small-scale technology solutions that generate energy from wastes. For example, the production of refuse-derived fuel using household and agricultural residues that are used to create briquettes as a source of domestic cooking fuel, or the production of biogas from agricultural residues wastes and sewage. In addition, for many African countries, the rates of waste collection by municipalities in the urban environment (towns and cities) are generally low and this has also provided some incentive for the private sector to develop in-house waste management solutions, some of which involve WtE. These solutions are mostly focused on the agro-processing; such as breweries, animal feedlots, and food processing industries.

CHAPTER FOUR

WtE Best Practices and Perspectives in Latin America

Suani Teixeira Coelho*, Daniel Hugo Bouille[†], Marina Yesica Recalde[‡]

*Research Group on Bioenergy, Institute of Energy and Environment, University of São Paulo, São Paulo, Brazil
[†]Fundación Bariloche, Río Negro, Argentina
[‡]Fundación Bariloche, Buenos Aires, Argentina

4.1 MSW MANAGEMENT AND POLICIES IN LATIN AMERICA

4.1.1 MSW Management and Policies in Argentina

Estela Santalla

Departamento Ingeniería Química, Facultad de Ingeniería/UNICEN, Buenos Aires, Argentina

In the framework of the National Law 26,190/2006 of Promotion of Renewable Energy for electricity production, and aiming to increase the share of renewable electricity, in 2009 the Argentinean Government launched the GENREN Program (an auction for electricity generation with renewable energies), with the objective of incorporate 1000 MW to the whole electricity market through the acquisition of electricity from renewable sources. When the program started, the amount of MW tendered was considered reasonable to put the country on track to achieve the 8% renewable energy goal in 2016 as stated in Law 26,190. In the first tender held in December 2009, a total of 895 MW was awarded in modules of up to 50 MW. The promotion policy through GENREN assured the suppliers selected by competitive bidding to purchase energy at a fixed price in dollars in the wholesale electricity market above market prices. However, even with the correct and efficient design and the awarding of a significant number of projects, works that represent less than 10% of the total awarded were completed. The reason for the failure of the GENREN Program is not found in the design of the standard, but rather it is the result of the international situation in which the access of Argentina to external financing was severely restricted (Aguilar, 2015).

The PROBIOMASA project[1] is an initiative of the ministries of Agro-industry and Energy and Mining, with the technical assistance of the Organization for Food and Agriculture (FAO), to promote the energy derived from biomass. The objectives are to increase production of thermal and electric energy derived from biomass at local,

---

[1] http://www.probiomasa.gob.ar.

Municipal Solid Waste Energy Conversion in Developing Countries
https://doi.org/10.1016/B978-0-12-813419-1.00004-8

provincial, and national levels to ensure an increasing supply of clean, reliable, and competitive energy, while opening new agroforestry opportunities, stimulate regional development and help mitigate the climate change. At the moment, the program contributes to evaluate the available biomass resources, advises public and/or private bioenergy projects to guarantee technical, economic-financial, social, and environmental sustainability and provides updated information and facilitates the management of financing opportunities. It also contributes to generate skills in workers, producers, technicians, university students, and the public on energy derived from biomass, sensitizes, and disseminates the benefits of biomass energy use, highlighting the care of the environment and its use as a renewable energy.

The Ministry of Science, Technology, and Productive Innovation promoted, through the National Plan of Science, Technology, and Innovation Argentina 2020,[2] the development of Strategic Productive Partner Cores assigning to each region of the country specific goals, such as the mitigation of GHGs through the recycling of different waste streams and the rational use of energy for Buenos Aires province. This action stimulated the formulation of private-public projects to develop the value chain to produce refuse derived fuels (RDFs) to be used, in principle, in the cement industry, with shared support of the official FONARSEC Sectorial Fund[3] and private companies. At the moment, two pilot projects of RDF production are under development.

In 2015, the Argentinean Government modified the existing renewable electricity law and enacted the National Law of Promotion of Renewable Energies (27,191/2015), in this framework a new renewable electricity auction was launched; and three rounds under the RenovAR Program were accomplished in 2018 with the adjudication of wind, photovoltaic, and biogas projects for electricity production. In 2018, a total of 35 MW for biogas projects, from a total power of 1200 MW, achieving an offer of only 8.6 MW from five biogas projects (swine wastewater, corn and vinasse silages, OFMSW) and one landfill gas (LFG) project were adjudicated. This scenario, although it shows signs of progress, is still far from being an established technology for the use of biogas for energy purposes.

4.1.1.1 Biodigestion in Argentina

Biodigestion is currently promoted in Argentina by the Government through the already mentioned PROBIOMASA. The main objective is to solve current and crucial environmental problems related to the inadequate management of agricultural residues such as manure in dairy, feedlots, and intensive breeding of pigs, added to a scenario of energy restrictions where the promotion of the biogas for electricity is framed in the current Law 27,191.[4]

---

[2] http://www.argentinainnovadora2020.mincyt.gob.ar/.

[3] It belongs to the National Agency of Scientific and Technologic Promotion (ANPCyT) http://www.agencia.mincyt.gob.ar/frontend/agencia/post/1775.

[4] National Law 27.191/2015 National Promotion Regime for the use of renewable energy sources for the production of electrical energy (Régimen de fomento nacional para el uso de fuentes renovables de energía destinada a la producción de energía eléctrica).

Also at national level, there is program Probiogas,[5] signed by the GEF project *Sustainable business models for the production of biogas from organic urban solid waste*. The Ministry of Environment and Sustainable Development and the United Nations Development Programme (UNDP) carry out this program. The objective is to demonstrate that biogas (from landfills and biodigesters) are technical, environmental, institutional, and economic sustainable for incorporate to the MSW management systems. This means that it can be operated by municipalities, and to support the national strategy to advance in the generation of renewable energy improving waste practices in the country. This project includes the construction of four pilot plants and technical studies to strengthen biogas production and use along 4 years.

4.1.1.2 Future Scenario for Biodigestion Development in Argentina

Public institutions such as the National Institute for Agricultural Technology (INTA), the National Institute for Industrial Technology (INTI), National Governmental Institutions such as the ministries of Agro-industry and Energy and Mining, and Universities are promoting the implementation of projects based on biodigestion in the agro-industrial sectors. A deeper synergy between them will be relevant for maximizing resources and provide integral solutions to technology users.

The Biolac network[6] promotes a sustainable territorial development and integration through the biogas, working up from 2009 sustained activities through courses, trainings, congress, and technical tours along several South-American countries such as Costa Rica, Nicaragua, Chile, Honduras, Colombia, México, Brazil, Peru, and Argentina. The main challenge of the organization is to improve the wellbeing of the Latin American and Caribbean people through the applied research and advocacy of biodigesters for the treatment and management of organic waste. Currently, Argentina actively participates in the network through the organization of congress, visits, and starts collaborative inter-laboratories studies for the biomethane potential determination in order to advance in the standardization of protocols.

PROBIOMASA, in the framework of the Promotion of Energy from Biomass Project jointly with the Energy Ministry and FAO, developed a spatial analysis of the energy balance from biomass in Argentina, applying the Wisdom methodology for eight provinces. Fig. 4.1 shows the current bioenergy projects developed in Argentina (left) and an example for the biogas potential generation for the province of Córdoba, where it can be observed the potential of biogas generation from dairy (right up) and swine (right down) productions.

The available bio-based resources in Argentina, gives to biodigestion technology huge opportunities to play a more significant role between renewable energies, in order to diversify the energy matrix to satisfy the growing demand of energy and at the same time to protect the environment.

---

[5] http://ambiente.gob.ar/gestion-integral-de-residuos/programa-probiogas/. Last access 20-01-2017.
[6] http://redbiolac.org/.

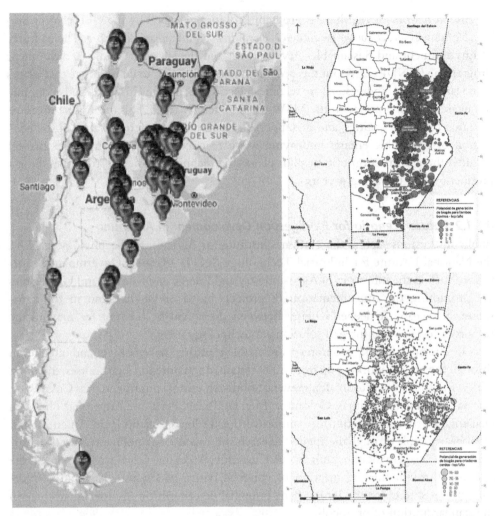

Fig. 4.1 Distribution of biomass projects in Argentina (2018). *(Source: http://www.probiomasa.gob.ar/ sitio/es/bp_mapa.php.)*

Technological solutions must be geared toward the distributed generation of energy, promoting energy self-sufficiency projects and industrial synergies for the valorization of industrial waste.

4.1.2 MSW Management and Policies in Brazil

Alessandro Sanches Pereira

Instituto 17, São Paulo, Brazil

Since solid waste and its impacts are associated with different areas of activity of the public power, many Brazilian legal instruments that deal with this issue are distributed in different governmental spheres. In addition to the National Solid Waste Policy (PNRS), there are various laws, decrees, resolutions, regulations, ordinances, and technical standards at the federal level that focus on the management and management of solid waste. Table 4.1 presents the timeline for the establishment of the PNRS (MMA, 2018).

The sanctioned Law incorporates modern concepts of solid waste management and is prepared to bring new tools to Brazilian environmental legislation such as (MMA, 2018):

- *Sectoral agreement*: an act of a contractual nature signed between the public authority and manufacturers, importers, distributors, or merchants, with a view to implementing shared responsibility for the product life cycle.

Table 4.1 Timeline of the Establishment of the National Solid Waste Policy in Brazil

Year	Description
1991	Draft Law 203 provides for packaging, collection, treatment, transportation, and disposal of health-care waste
1999	Proposition CONAMA[a] 259 entitled Technical Guidelines for the Management of Solid Waste. Approved by the council plenary, but not published
2001	Chamber of Deputies creates and implements the Special Committee on National Waste Policy with the objective of assessing the matters contemplated in the bills attached to Draft Law 203/91 and formulating a global substitute proposal. With the closure of the legislature, the Commission was extinguished
	The 1st National Congress of Recyclable Material Collectors was held in Brasilia, with 1600 congressmen, among collectors, technicians, and social agents from 17 states. They promoted the 1st National March of the Street Population, with 3000 participants
2003	In January, the First Latin American Congress of Waste Pickers was held in Caxias do Sul, which proposes professional training, waste disposal eradication, and waste generator accountability
	President Lula institutes the Inter-ministerial Work Group on Environmental Sanitation in order to promote the integration of environmental sanitation actions within the federal government. GT restructures the sanitation sector and results in the creation of the Urban Solid Waste Program
	The 1st Environmental Conference was held
2004	The Brazilian Environmental Ministry (MMA) promotes inter-ministerial discussion groups and ministry secretariats to preparation of a proposal for the regulation of solid waste
	In August of the same year, CONAMA held a seminar entitled "Contributions to the National Policy on Solid Waste" with the objective of listening to society and formulating a new proposal for a bill, as the CONAMA 259 proposal was out of date
2005	Created an internal group in the Secretariat of Environmental Quality in MMA Human Settlements to consolidate Conama Seminar contributions, draft legislation in the National Congress and the contributions of the various actors involved in solid waste management
	A draft "National Solid Waste Policy," Ministries of Cities and Health, through its National Health Foundation-Funasa, Development, Industry and Foreign Trade, Planning, Budget and Management, Social Development and Fight against Hunger and the Treasury
	The 2nd National Conference on the Environment was held to consolidate the participation of society in the formulation of environmental policies. One of the priority themes is solid waste
	Regional solid waste seminars were promoted by CONAMA, Ministry of the Environment, Ministry of Cities, Funasa, Caixa Econômica Federal and debates with the National Confederation of Industries (CNI), Federation of Industries of the State of São Paulo (FIESP), Brazilian Association of Sanitation Engineering (ABES), Corporate Commitment

Continued

Table 4.1 Timeline of the Establishment of the National Solid Waste Policy in Brazil—cont'd

Year **Description**

for Recycling (CEMPRE), and other entities and related organizations, such as the Garbage & Citizenship Forum and the Inter-ministerial Committee on Social Inclusion of Garbage Collectors

New Special Commission has been set up in the Chamber of Deputies

2006 Approved report dealing with PL 203/91 plus the release of the importation of used tires in Brazil

2007 In September, the Executive Board proposes PL 1991. The draft Law on the National Solid Waste Policy considered the lifestyle of contemporary society, which together with the marketing strategies of the productive sector, lead to intensive consumption, causing a series of impacts environmental, public and social health incompatible with the sustainable development model that is intended to be implemented in Brazil

Proposed Law 1991/2007 presents a strong interrelationship with other legal instruments at the federal level, such as the Basic Sanitation Law (Law 11,455/2007) and the Public Consortia Law (Law 11,107/1995), and its Regulatory Decree (Decree No. 6.017/2007). Likewise, it is interrelated with the National Environmental, Environmental Education, Water Resources, Health, Urban, Industrial, Technological and Foreign Trade Policies and those that promote social inclusion

Text is finalized and sent to the Civil House

Constituted WG (GTRESID) to analyze substitute sub-proposal proposed by the rapporteur, which involved meetings with the Civil House

2008 Public hearings were held, with contributions from CNI, representatives from interested sectors, the National Movement of Recyclable Material Collectors, and other members of GTRESID

2009 In June, a draft of the Final Report was submitted to receive additional contributions

2010 On March 11, the plenary of the Chamber of Deputies approved, in a symbolic vote, a substitute for Senate Bill 203/91, which establishes the National Solid Waste Policy and imposes obligations on entrepreneurs, governments, and citizens in the management waste

Then the bill went to the Senate. It was examined in four committees and on July 7 was approved in plenary

On August 2, President Luiz Inacio Lula da Silva sanctioned the law creating the National Policy on Solid Waste (PNRS)

On August 3, it is published in the Official Gazette of the Union Law No. 12,305 that establishes the National Solid Waste Policy and provides other measures

On December 23, Decree No. 7404, which regulates Law No. 12,305, of August 2, 2010, which establishes the National Solid Waste Policy, is published in the Official Gazette of the Union and creates the Inter-ministerial Committee of the National Solid Waste Policy and the Steering Committee for the Implementation of Reverse Logistics Systems, and makes other arrangements

Also on December 23, Decree No. 7405, which establishes the Pro-Garbage Pickers Program (i.e., Programa Pró-Catador), is called the Inter-ministerial Committee for Social and Economic Inclusion of Waste Disposers and Recyclables. The Inter-ministerial Committee on Social Inclusion of Waste Disposers created by the Decree of September 11, 2003, provides for its organization and operation, and makes other arrangements

[a]National Environment Council (CONAMA).

Source: MMA, 2018. Política Nacional de Resíduos Sólidos. https://www.mma.gov.br/pol%C3%ADtica-de-res%C3%ADduos-s%C3%B3lidos. (Accessed 29 August 2019).

- *Responsibility shared by the product life cycle*: set of attributions of manufacturers, importers, distributors and traders, consumers, and holders of public services of urban cleaning and solid waste management by minimizing the volume of solid wastes and tailings generated, as well as by reducing the human health and environmental quality impacts resulting from the product life cycle, under the terms of this Law.
- *Reverse logistics*: an instrument of economic and social development, characterized by a set of actions, procedures, and means to enable the collection and restitution of solid waste to the business sector, for reuse in its cycle or in other production cycles, or other destination environmentally sound end.
- *Selective collection*: collection of solid waste previously segregated according to its constitution or composition.
- *Product life cycle*: a series of steps that involve the development of the product, the obtaining of raw materials and inputs, the production process, the consumption, and the final disposal.
- *Solid waste management information system—SINIR*: aims to store, process, and provide information that supports the functions or processes of an organization. Essentially it is composed of a subsystem composed of people, processes, information, and documents, and another composed of equipment and its means of communication.
- *Collectors of recyclable materials*: several articles approach the theme, with the incentive to mechanisms that strengthen the performance of associations or cooperatives, which is fundamental in solid waste management.
- *Solid waste plans*: The National Solid Waste Plan to be developed with broad social participation, containing national targets and strategies on the theme. State, micro-regional, metropolitan, inter-municipal, municipal solid waste management plans, and solid waste management plans are also foreseen.

4.1.3 MSW Management and Policies in Ecuador

Rafael Soria and Laura Salgado

Departamento de Ingeniería Mecánica, Escuela Politécnica Nacional, Ladrón de Guevara, Quito, Ecuador

In Ecuador there is not a specific law regarding waste to energy. However, there are policy instruments that provide guidelines and technical normative and promote the implementation of integrated waste management systems. Following a legal hierarchy, the most important instruments are the National Constitution, the Organic Code for Territorial Autonomy and Decentralization, the Organic Environmental Code, and the Law of Environmental Management.

The National Constitution of Ecuador was launched in 2008 and it was one of the first constitutions that recognize Rights of Nature. As the main legislation instrument, it indicates that the State must promote the use of environmentally friendly technologies and alternative energy sources with low environmental impact (Article 15 and Article 413).

Article 254 clause 4 states that the municipal governments (Decentralized Autonomous Governments—GADs) must provide public services such as waste disposal and collection. According to article 415, municipalities are also in charge of implementing programs and projects regarding water usage and management of solid waste and sewage water (Asamblea Constituyente, 2008).

4.1.3.1 National Law for Territorial Autonomy and Decentralization
This instrument was established in 2010 and regulates the obligations and attributions of Local Governments. Article 8 clause (a) establishes that municipalities are autonomous to develop and implement regulation or other policy instruments to support their work within their jurisdiction. Article 57 states that municipalities have the capacity to create public or mix companies to provide public services that are part of their attributions. It was established that Municipal Governments should progressively implement integrated systems for waste management (article 136) (Asamblea Nacional, 2010).

4.1.3.2 Organic Environmental Code and Secondary Legislation of Environment
The Organic Environmental Code published in 2017 presents the Title V "Integrated Waste Management" which includes a complete list and description of general policies and principles in this field (Article 225) (Asamblea Nacional, 2017). Here the importance of waste recovery and recycling is highlighted as well as new topics that have not been presented before in the national legislation such as social inclusion in waste recovery activities and the extended producer responsibility (EPR, Articles 226–233). In contrast to the Organic Environmental Code, the Secondary Legislation of Environment provides through Folio VI "Environmental Quality," Chapter VI about "Integrated Waste Management," a more detailed definition of principles and technical standards about the remediation techniques for open dumps, phases for an integrated waste management system, rules for environmental licences, and others. However, the specifications and scope of this policy instrument does not present incentives or technical guidelines for waste-to-energy uses (Ministerio del Ambiente, 2015).

4.1.4 MSW Management and Policies in Mexico
Gustavo Solórzano

Asociación Mexicana de Ingeniería, Ciencia y Gestión Ambiental, A.C. (AMICA), Mexico City, Mexico

4.1.4.1 Waste and Environment
The legal framework relating to the environment formally began in Mexico in 1971, as an antecedent to the *United Nations Conference on the Human Environment* which took place in Stockholm, Sweden, in 1972, although some isolated legal instruments previously existed. The *Federal Act to Prevent and control pollution* was published in 1971, and later it was repealed and replaced in 1982 by the *Federal Act for Environmental Protection*. In turn this instrument was replaced by the *General Act for Ecological Equilibrium and Environmental Protection* (known

as LGEEPA, for its acronym in Spanish) in 1988, which continues to be in force in 2018. It is a comprehensive measure, since it regulates such diverse fields as air, water, soils, waste, natural resources, etc. In order to facilitate the application of these legal instruments, each one includes thematic regulations or rules, as well as Official Mexican Standards or norms.

Although the LGEEPA includes a specific section focused on waste (hazardous and nonhazardous), in 2003, the *General Act for the Prevention and Integrated Management of Waste* (LGPGIR, for its acronym in Spanish) was published with the aim of more efficiently regulating three specific categories of waste: urban solid waste, hazardous waste, and special handling waste (e.g., wastewater treatment sludges, construction and demolition waste, tires, etc.). Since it was passed, the LGPGIR has been slightly modified on several occasions, but it currently requires a thorough review due to the fact that it only mentions marginal aspects related to the use of energy from waste. As it stands, this act has its specific regulatory instrument and also delineates an important number of official standards or norms that serve diverse particularities such as those relating to the design, construction, and operation of landfills; safety confinement for hazardous waste; and atmospheric emissions from waste incineration plants, waste coprocessing in cement kilns, among others.

An additional legal instrument is the *General Act for Climate Change* posted in 2012, which, among other recommendations includes clear and explicit guidelines for the use of energy from the different categories of waste.

In the field of policy instruments on waste, the *Policy and strategies for the integrated waste management* was published in Mexico in 2007. Derived from this document, two *National programs for the management of solid and hazardous waste* were published, first in 2009 and second in 2017. Based on the 2007 document, it was established that each state and municipality in Mexico should develop its own program for the prevention and management of waste. It is interesting to note that, although Mexico is a member of the Organization for Economic Cooperation and Development (OECD), no legal or policy instrument in Mexico contemplates or mentions the concept of the producer's extended responsibility, focusing instead on the shared responsibility of all the stakeholders involved.

As regards the organic structure of the Federal Government in Mexico, the *Ministry for Environment and Natural Resources* (known as SEMARNAT, for its acronym in Spanish) is responsible for the design of public policies in the field of the environment, including waste management. There is also the *National Institute for Ecology and Climate Change* (INECC, its acronym in Spanish), which acts as the technical and scientific branch of SEMARNAT. Finally, the Attorney General's office or *Federal Agency for the protection of the environment* (PROFEPA, its acronym in Spanish) is the entity of the Federal Government responsible for law enforcement on national environmental matters.

4.1.4.2 Energy

Traditionally, the generation, transmission, and commercialization of all types of energy in Mexico was limited and controlled by a state-owned company. However, in 2013, an energy reform began the most outstanding aspects of which are three: opening up the possibility of private participation in the energy market; the formal incorporation of renewable energies into the energy market; and the establishment of clean energy certificates. Since then, various entities of public administration are in the process of elaborating the necessary legal and policy instruments, as well as the infrastructure to implement and meet the objectives of the aforementioned reform.

The new legal instruments are represented by the *Act on Electric Industry*; the *Act on the Promotion and Development of Biofuels*, which contemplates biogas as biofuel; and the *Act for Energy Transition*. As one special policy instrument, among others, the *National Energy Strategy 2013–2027* was published in 2013. All of these instruments explicitly incorporate the use of energy from waste, and wind, geothermal, hydro, mini hydraulic, biomass, and solar energies are also mentioned. It is important to mention that in Mexico waste incineration is considered a clean energy activity, provided that the waste-to-energy facilities comply with the applicable environmental standards, in particular those relating to atmospheric emissions and other types of discharges to the environment.

Within the organizational structure of the National Government related to the use of waste as a source of energy, the main entities are represented by the *Energy Ministry* (known as SENER, its acronym in Spanish), the *Energy Regulatory Commission* (CRE, its acronym in Spanish), and the *National Energy Control Centre* (CENACE, its acronym in Spanish).

4.2 ENERGY ACCESS IN LATIN AMERICA: CURRENT SITUATION AND DIFFICULTIES TO INCREASE ACCESS

Marina Yesica Recalde
Fundación Bariloche, Buenos Aires, Argentina

The role of energy consumption in socioeconomic development has been widely discussed from theoretical and empirical standpoints. According to some authors, energy constitutes a biophysical constraint to economic growth because every activity requires a minimum quantity of energy to be performed (i.e., Reister, 1987; Tahvonen and Salo, 2001; Recalde and Ramos-Martin, 2012; among others). Socioeconomic development requires a minimum of energy to be properly delivered (Recalde, 2017). The contribution of energy to sustainable development dimensions (economic, political, social, and environmental) is straightforward.

This relation of energy and human development is highly related to the definition of energy poverty, a discussion that boosted in the literature in recent decades. As stated by Bhatia and Angelou (2015), energy poverty can be defined as "the state of being deprived of certain energy services or not being able to use them in a healthy, convenient, and efficient manner, resulting in a level of energy consumption that is insufficient to support social and economic development." There is, therefore, a clear link between energy poverty and energy access.

International community has also highlighted the link between energy and socioeconomic development by the recognition that it will not be possible to achieve the Sustainable Development Goals (SDGs) without an increase in global energy access. Access to appropriate, adequate, and affordable energy is closely linked to achievement of several developmental objectives, including gender equality, education, health, food security, rural development, and poverty reduction (Bhatia and Angelou, 2015). According to the BID, in the LAC region 4,3 MM of persons per year die because of different diseases related to air pollution in households due to inefficient burn of wood, charcoal, and other fuels. In this regard, as stated by United Nations, promoting sustainable energy is an opportunity to transform lives, economies, and the planet. Therefore, the seventh SDG states "Ensure access to affordable, reliable, sustainable and modern energy for all." In this regard, UN has defined four different targets, and the corresponding indicators, that will be useful to evaluate the performance of the regions and the world toward the SDG7, these are shown in Table 4.2; targets and indicators directly related to energy access have been highlighted.

According to the progress toward the SDG Reports of the Secretary-General, in 2017 related to the energy area seem to fall short of what is needed to achieve energy access for all and to meet targets for renewable energy and energy efficiency. In particular, according to this report, there has been a little progress in access to electricity and access to clean fuels and technologies for cooking between 2012 and 2014. The report also highlights the need for higher levels of financing and bolder policy commitments.[7]

It is important to note that the idea of "modern energy for all" needs to be deeply analyzed as the situation of energy access may be very different from developing regions and countries and within them. For instance, although LAC can be seen as the most advanced region among developing ones in terms of access to energy, the situation is very dissimilar among countries considering that the LAC region is very divergent in socioeconomic terms and, therefore, there is still a lot of work to do. According to information from OLADE, nearly 40 MM out of the total population of 600 MM of persons in LAC lack of access to modern energy. For instance, while 97% of inhabitants have access to electricity in many countries, several individual countries have high shares of people without access, including Haiti (67% of the population; 7 million people), Honduras (24%; 2 million), and Nicaragua (11%; 0.7 million). On the other hand, the countries studied in this book have high electrification rates, such as Argentina, (95%), Ecuador (97%), Brazil, (98), and México (99%).

One of the most relevant challenges in measuring access to energy is the absence of a universal definition of the concept of access. Therefore, any discussion on access to energy should begin with a discussion and clear definition of the concept of access to energy itself. As stated by the Bhatia and Angelou (2015), the lack of a universal definition of this concept resulted in several approaches attempting to measure access to energy

---

[7] http://www.un.org/ga/search/view_doc.asp?symbol=E/2017/66&Lang=E.

Table 4.2 SDG7 Targets and Indicators

Target	Indicator
7.1 By 2030, ensure *universal access to affordable, reliable, and modern energy services*	**7.1.1** Proportion of *population with access to electricity* **7.1.2** Proportion of *population with primary reliance on clean fuels and technology*
7.2 By 2030, increase substantially the share of renewable energy in the global energy mix	**7.2.1** Renewable energy share in the total final energy consumption
7.3 By 2030, double the global rate of improvement in energy efficiency	**7.3.1** Energy intensity measured in terms of primary energy and GDP
7.A By 2030, *enhance international cooperation to facilitate access to clean energy research and technology,* including renewable energy, energy efficiency and advanced and cleaner fossil-fuel technology, and promote investment in energy infrastructure and clean energy technology	**7.A.1** Mobilized amount of United States dollars per year starting in 2020 accountable toward the $100 billion commitment
7.B By 2030, *expand infrastructure and upgrade technology for supplying modern and sustainable energy services for all in developing countries, in particular least developed countries, small island developing States, and land-locked developing countries,* in accordance with their respective programs of support	**7.B.1** Investments in energy efficiency as a percentage of GDP and the amount of foreign direct investment in financial transfer for infrastructure and technology to sustainable development services

Source: Inter-Agency and Expert Group on SDG Indicators (IAEG-SDGs).

using a variety of indicators. In this sense, this concept is usually associated to household access to electricity. Nevertheless, as stated by the World Bank in the Energy Sector Management Assistance Program (ESMAP) document prepared for the SE4all project, this definition ignores energy for cooking and heating needs, as well as for productive engagements and community facilities, and it does not address the questions of affordability of energy and legality of connection.

The analysis of energy access must be developed in this direction. Affordability of energy is crucial to analyze energy access, and it is important to note that affordability is not a unique variable depending; instead of that, affordability of energy is a function of the package of energy, the price of energy, and the user's income level (Bhatia and Angelou, 2015). As clearly stated by Pachauri and Spreng (2003) the physical access to energy end-use equipment is a prerequisite for access to energy services, but it does not assure access to energy services. Real access to energy services can be limited by the purchasing power of the household, the cost of energy, and cost of equipment. In this sense, the key variables to understand affordability are the market prices of the energy sources, efficiencies of energy sources, and costs and efficiencies of appliances needed for specific energy fuels.

It is important to note that the improvement in energy access is not a single-step transition from lack of access to availability of access. Instead of this, access refers to a path to increasing levels of energy attributes (Bhatia and Angelou, 2015). This aspect includes not only the access to modern and clean energy sources, but also the affordability of energy.

There are different ways to measure energy access in order to monitor and evaluate the improvements. In the first place, and within the most common and simplify approaches, there are the binary metrics, that rely on a single minimum threshold of energy supply or services to determine the number of households that can be considered as having access, and it can be defined as use of certain fuels, electricity grid connection, among others. The con of this approach is that it ignores the quality of energy accessed and other aspects of energy use. In the second place, the UNDP/WHO Dashboard of Indicators presents data as a series of percentages of population with access to specific energy supplies and equipment. It includes information on percentage of people with a household electricity connection; percentage of people who use electricity/liquid fuels/gaseous fuels as their primary fuel to satisfy their cooking needs; percentage of people who use different types of cooking fuels as their primary cooking fuel, including both modern and solid forms of energy; among other indicators. In the third place, there is the energy poverty line, which requires the definition of a minimum level of energy to satisfy basic human needs. The cons of this approach consist in the fact that minimum energy need itself varies depending on climatic conditions, cultural preferences, economic conditions, among other factors. An improved version of this index is the Multidimensional Energy Poverty Index (MEPI) developed by UNIDO that measures energy poverty and report progress of energy access policies, based on the recognition of the multi-dimensional nature of energy poverty and the need to capture nexus between access to modern energy services and human development (Nussbaumer et al., 2012). In the fourth place, a new an interesting composite index to measure country evolutions to the use of sustainable energy is the Energy Development Index (EDI) of the International Energy Agency, which focus on four indicators: commercial energy consumption per capita, electricity consumption in residential sector per capita, share of modern fuels in total residential sector energy use, and share of population with access to electricity. More recently, the World Bank in the SE4All project has defined an Overall Energy Access Index, which includes an Index of Household Energy Access, composed by three different indexes: electricity, cooking, and heating. The idea of this integrated index is to include the main energy services necessary for the contribution of energy to sustainable development mentioned before and it is defined based on a group of attributes of each energy services and multitiers. One of the key challenges identified and highlighted by the SE4All project refers to the availability of data in existing households' surveys. In the case of Latin American region, some countries are currently conducting household energy surveys by their own agencies, this is, for example, the case of Argentina. However, in terms of the evaluation of the performance of the countries, it will be important to have comparable and standardized surveys periodically developed.

4.2.1 Electricity Access in Latin American Region

As mentioned previously, it would be more appropriate to measure energy access based on integrated index. However, data availability to compare different countries and regions is a barrier for this type of analysis. Therefore, most energy access studies rely in the aforementioned binary metrics and they are used in this section to evaluate the electricity access of Latin American region. Figs. 4.2–4.4 represent the situation of total, rural, and urban electricity access in the region and selected countries.

As stated by Coviello and Ruchansky (2017) at least the last two or three decades have been characterized by a progress in the rate of electricity access. This, somehow, can be an indicator of the performance of the public policies to enhance these services. The Latin American region, except a small group of countries, has been characterized by a quite slow progress in 2012–14 with annual access growth rate between 0 and 2 percentage point, and only a couple of countries increasing the electrification rates by 2–3 percentage points annually.

Nevertheless, despite the recent progress in electrification in 2016 there were still 26 MM people without access to electricity in the region (4% of total population). Although nearly 97% out of total population has access to electricity, there is a clear difference between electricity access in rural and urban regions. Coviello and Ruchansky (2017) show that while almost all urban population has electricity access, there is a lot of rural population without such access, mainly due to the difficulties to accessing rural

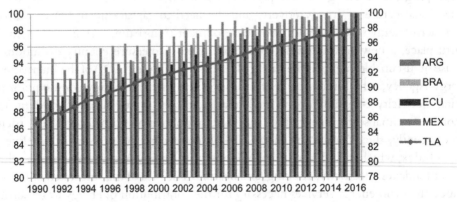

Fig. 4.2 Electricity access (%).

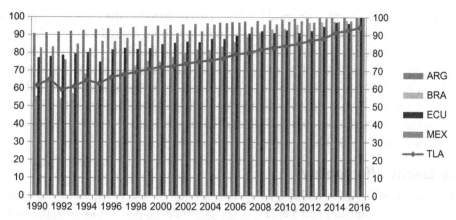

Fig. 4.3 Rural electricity access. % of total rural population.

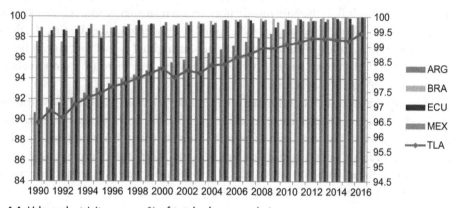

Fig. 4.4 Urban electricity access. % of total urban population.

areas and the higher costs of rural electrification. However, according to the authors, LAC region is the developing region of the world with the smallest gap between urban and rural.

Regarding the countries analyzed in this book, it is important to mention that Argentina, Brazil, and Mexico are within the most important countries of the region regarding the GDP and energy consumption, and that they have very high electrification rates, which has impact on the average electrification rate of the region.

4.2.2 Access to Clean Cooking Technologies and Fuels

According to BID, in 2016 87 MM people rely on traditional solid fuels (wood and charcoal) for cooking, and nearly 2800 MM do not have the most appropriate cooking installations, with the corresponding health problems mentioned before.

In this regard, the 2000–14 period has been characterized in the region by a growing trend to the access to modern cooking fuels, although there is a reduction in the rate of growth of the penetration of these technologies in the last years (Coviello and Ruchansky, 2017). In 2016, 87% of regional population had access to clean ways of cooking, a situation that is clearly much more promising in Argentina, Brazil, and Ecuador than in México (Fig. 4.5). As argued by Coviello and Ruchansky (2017), if the current trend is projected, then nearly all LA population may have access to clean energy sources for cooking in 2030.

4.2.3 Energy Affordability

As mentioned before, affordability is a key for analyzing the electricity and energy access. For the LA region, Jiménez and Yépez-García (2017) developed a cross-sectorial analysis based on energy expenditures of 13 countries of the region. One important aspect highlighted by the authors is that in most of the cases, the traditional energy sources have low impact on energy bill, as they are mostly noncommercial energy sources. As expected the share of energy bill on total household expenditures is higher for the first deciles of income (poorest households). Additionally, the report highlights that for poorest households of the region electricity has the highest energy budget share (4.5% on average), followed by natural gas (3% average).

Considering the relevance of affordability for energy access, National Governments use to develop energy policies to reduce energy costs in poorest regions and, then, increase energy access. This has been indeed the case for many LA countries, in which there has been a recent relevant effort to facilitate the access toe energy equipment of modern energy sources, based on subsidies and social tariff schemes among other instruments of policy (Coviello and Ruchansky, 2017).

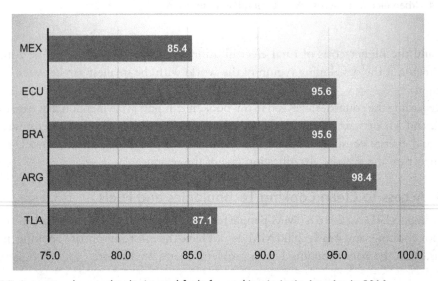

Fig. 4.5 Access to clean technologies and fuels for cooking in Latin America in 2016.

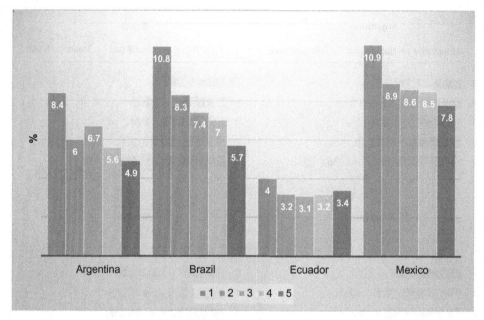

Fig. 4.6 Energy expenditure as share of total expenditure in different LA countries.

Fig. 4.6 shows the energy expenditure as share of total expenditure in the four LA countries that have been selected as case of studies, and Fig. 4.7 shows the expenditure in different energy sources: Electricity (first bar), natural gas (second bar), and transport fuels (last bar) for the same countries. There are some points that need to be highlighted. Firstly, in all the countries in the poorest quintile the share of energy expenditures is higher than the highest, being Argentina and Brazil the most inequitable cases: in Argentina while the first quintile energy represents 8.4% of its total expenditure, the first one expends 4.9%, in Brazil the difference is 10.8% and 5.7% and in México 10.9% and 7.8% (see Fig. 4.6). Secondly, the situation seems to be more inequitable for some specific fuels, as shown in Fig. 4.7, this is the case of electricity in Argentina (3% against 0.8% out of total expenditures), Brazil (5.3% and 1.8%), and Mexico (3.5% and 1.4%), and less inequitable in Ecuador (2.4% and 1.5%); and the natural gas. In the case of transport fuels, as expected the share of expenditures is bigger in the highest quintiles than in the lowest one. This aspect has been largely discussed by Yépez-García (2017).

4.2.4 Energy Access and Rural Electrification in Brazil

Osvaldo Soliano Pereira

Universidade Federal da Bahia—UFBA (Federal University of Bahia), Salvador, Brazil

Rural electrification in Brazil dates back to the 1970s, when several programs were implemented to increase the penetration of electricity to country's rural regions and

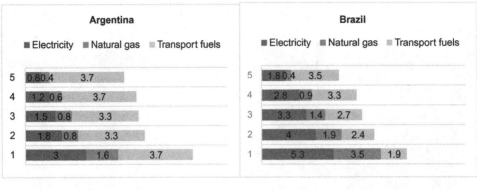

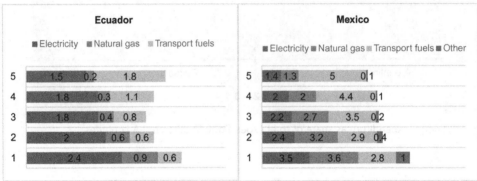

Fig. 4.7 Energy expenditure as share of total expenditure by fuels in different LA countries.

thereby encourage their socioeconomic development. Loans from multilateral agencies, such as the World Bank and the Inter-American Development Bank used to finance expansion programs until mid-1980s.

Impact assessment studies conducted throughout the late 1980s showed that there was a disconnection between the expected impacts of rural electrification and the results achieved from both the economic and social viewpoints (Soliano Pereira, 1992). The moment coincided with the crisis of the 1980s. The result was a disruption in the flow of international funding and halting electrification programs.

In the early 1990s, the consolidation of the concept of sustainable development expanded the expectations from rural electrification, introducing the need to modern fuels and universal access to power services.

The low level of rural service throughout the 1990s prompted the Brazilian Government to establish the National Rural Electrification Program "Luz no Campo" (PLC) in 1999. It aimed to electrify 1 million rural properties/households until 2003, through interconnection to national grid. The PLC ended up interconnecting approximately 735,000 by the end of horizon of the program. Despite falling short of the original target, their results created a new dynamic for a segment of the electricity sector that was coming at a very slow pace.

The enactment of Law No. 10,438, of April 26, 2002, was undoubtedly the highest point of universalization in the country. The Law established the compulsory nature of the universal service of electricity, established that the National Electricity Agency (ANEEL) should set targets and deadlines for this, ensured that the service of new consumers should be free of charge and secured resources with the Global Reversion Reserve (RGR)—soft loans—and the Energy Development Account (CDE)—grants. By 2003, ANEEL defined the universalization goals for each municipality in the country and established the national limit for full attendance in 2015.

Although Law and Resolution established the obligation, criteria, and deadlines for universalization, it failed to consider how to maintain the economic-financial balance of concession contracts and to ensure the adequate provision of services—regularity and continuity. Technical difficulties and the cost of providing services within each concession area determined the deadlines for effecting service, and thus, given its more adverse location and low demand expectation, the poorest population tended to be considered at the end of the deadline stipulated by ANEEL. These difficulties were visible when concessionaires presented their initial universalization plans by 2004.

In order to take over these constraints Federal Government created, in 2004, a national program, called Light for All (LpT), making possible to implement and accelerate the legal obligation, regulated and supervised by the regulatory agency. The purpose of LpT Program was to anticipate the goals, through adequate institutional and financial mechanisms, and to promote integrated local/rural development actions.

Main objective of the program was the mitigation of the electric exclusion that was centered, in more than 80%, in the rural area. Although clearly unlikely, LpT has set itself the goal of achieving its objectives by the end of 2008, 7 years ahead of the deadlines originally regulated by ANEEL. Some problems postponed the achievement of unrealistic deadline initially established, such as, unreliable available data on unattended population and the volume of resources required, substantially larger than initial estimations. Main consequence of latter problem was the impact on the tariffs, which ended up becoming a key obstacle to the rapid implementation of universalization.

A major advance of the LpT was the assembled arrangement, which involved sectoral funds, such as the RGR and the CDE, instead of using treasury resources, and the state and municipal partnerships. The map of the country's electricity exclusion showed that families without access to energy were mostly living in rural areas of municipalities with lower Human Development Index and low-income families.

An important aspect to note is that LpT has established service alternatives beyond the conventional grid extension by assuming that decentralized generation systems, either small local grids or individual systems, were acceptable to fulfill universalization targets. An ANEEL Resolution, from 2005, regulated the procedures and conditions of supply through Individual Generation Power Systems with Intermittent Sources of Energy (SIGFI) by distribution concessionaires.

In 2008, Government extended LpT's deadline to 2010, not only considering implementation problems in some states, but also the need to broaden state goals. In mid-2011, Government extended the Program again, with the expectation to be completed in 2014, 7 years after the initial forecast, and much more in line with ANEEL's initial forecast. Thus, ANEEL approved revision of the universalization plans. Once again, in 2014, Government extended LpT until 2018, deadline that is not enough to universalize the service in Brazil.

According to Eletrobras, under the LpT, the end of 2016 made 2.7 million connections, and the end of the Program foresees another 347 thousand. In total, around 16 million people will benefit.

Data from the National Survey of Household Sample (PNAD) (IBGE), referring to the year 2015, indicated that there were 197 thousand households remaining without access to electric lighting, which represented 0.3% of the country's permanent private households. However, these numbers are much lower than those made available by distribution concessionaires in their Plans to Universalization. According to ANEEL (2018), those Plans forecast the connection almost 107 thousand new consumers between 2019 and 2021, beyond the deadline of LpT, which is expected to be extended once again. These figures are not definitive as final numbers concerning 2017 are not available yet, and those forecast to 2018 hardly will happen due to recent economic crises faced by Brazil, which affected to power sector. A small part of the connections forecast to be installed between 2017 and 2018 makes use of generation systems not connected to the grid, 25.3 thousand out nearly 220 thousand.

The state of Pará will be the last one to have universal electricity service, which will happen by 2022, although there is no definition of how many systems will be installed from 2019. According ANEEL, Bahia will have the service universalized by 2021, with almost 24 thousand households to be connected per year between 2019 and 2021. The timetable for the concession areas, where it has not been achieved yet, has been approved by ANEEL through 2015–17. In some case, they will be reviewed, due to non-accomplishment of the targets, particularly some federal concessionaries on the brink to be privatized.

4.2.4.1 Challenges for the Future of Universalization

The challenge of universalizing the electric power service is still far from complete. Certainly, it will not remain in the 106 thousand consumers planned by ANEEL for the 2019–21 horizons.

Without the LpT, from 2019 onwards, the deadlines for achieving full coverage are to be expanded. It is hard to imagine that a company like COELBA, in the state of Bahia, will run 24 thousand new connections per year in the rural area, with only its own resources. Between the dilemma of raising the energy tariff or postponing universal access, government and electric sector will probably opt for the latter. Drought that hit the country's hydroelectric reservoirs increasing the dispatch of thermoelectric plants, which are much more expensive, has resulted into significant increases in final tariffs. On the other side, the two consecutive years of recession (2015/16) emptied the coffers of national and state treasuries, what reduced substantially the resources to accelerate full coverage.

The eventual closure of LpT in 2018 could aggravate the situation of electric exclusion in rural areas. An alternative that could certainly contribute to minimizing the impact on the volume of resources, which is already perfectly regulated, is the use of SIGFI and, more recently, the MIGFI (Isolated Microsystem for Generation and Distribution of Electricity). They are still little encouraged have a very limited use in the country. MIGFI could use a plethora of diverse technologies including the energy produced with residues, which would contribute to minimize a chronic problem of the Country that is the final disposition of solid residues. This possibility could leverage of the necessary resources with the involvement of municipal governments and collection and disposal companies.

Recent regulation on micro and mini power producers allowed energy compensation between the consumer and the concessionaire in 2012, enabled the creation of condominiums, consortia, and energy cooperatives in 2015. These advances made possible the integration of generation systems, originally individual or small grids, to regional or national grid as micro and mini power "prosumers." Waste energy can therefore serve an isolated area of the network or a group of consortium consumers connected to grid.

The challenge of universalizing persists throughout the society, with a role for the government, providing resources, for the regulatory agency, seeking means to encourage diversification and sources and minimizing the impact on future tariffs, and for entrepreneurs, by the identification of new business opportunities. Acting together, these actors can accelerate the prospects that all Brazilians will have access to electricity until the beginning of the next decade.

4.3 WtE TO INCREASE ENERGY ACCESS IN LATIN AMERICA

Daniel Hugo Bouille and Suani Teixeira Coelho

Fundación Bariloche, Río Negro, Argentina
Research Group on Bioenergy, Institute of Energy and Environment, University of São Paulo, São Paulo, Brazil

As mentioned before, SDG7 has a clear goal for the energy system, to ensure access to affordable, reliable, sustainable, and modern energy for all. However, nearly 1 billion people of the world's population do not have access to electricity. The health and well-being of some 3 billion people are adversely impacted by the lack of clean cooking fuels, such as wood, charcoal, dung and coal, which causes indoor air pollution.

WtE technologies can play a role on a major and better access to energy services and to contribute to SDG7: "Our everyday lives depend on reliable and affordable energy services to function smoothly and to develop equitably. A well-established energy system supports all sectors: from businesses, medicine and education to agriculture, infrastructure, communications and high-technology. Conversely, lack of access to energy supplies and transformation systems is a constraint to human and economic development."

Providing access to improved energy carriers is clearly a necessary, but insufficient condition for overall poverty alleviation and socioeconomic growth. Alleviating poverty, in its totality, clearly also requires improving the earnings of the poor by providing them with more sustainable livelihood opportunities through encouraging the use of energy in activities that can generate income. This requires defining access in a much broader sense and would require making available reliable and adequate qualities and quantities of energy and the associated technologies at affordable costs in a manner that is socially acceptable and environmentally sound so as to meet basic human needs and for activities that are income generating and could empower growth and development. Such a broader definition of access includes several elements and dimensions, including quality, reliability, adequacy, affordability, acceptability, and environmental soundness. Unfortunately, national level indicators and statistics to measure and monitor these various dimensions of access are extremely scarce, particularly for the least developed countries and regions where the issue is the most pressing. In Latin America, North Africa, the Middle East, and East Asia, the pace of electrification outstripped the rate of growth of the population by a large margin, so that access significantly improved.

The amount of MSW, one of the most important by-products of an urban lifestyle, is growing even faster than the rate of urbanization. Municipal solid waste management is the most important service a city provides; in low-income countries as well as many middle-income countries, MSW is the largest single budget item for cities and one of the largest employers. Solid waste is usually the one service that falls completely within the local government's purview. A city that cannot effectively manage its waste is rarely able to manage more complex services such as health, education, or transportation.

The most recent prospective is the WEC Report from 2016 remarks: Biological WtE technologies will experience faster growth at an average of 9.7% per annum, as new technologies (e.g., anaerobic digestion) become commercially viable and penetrate the market. In any case the faster growth will be in Asia-Pacific Region. WtE market will continue to develop globally as governments will impose supportive regulation with subsidies and tax benefits. One of the biggest barriers to market development will be the high technology costs in comparison with landfilling, which is the most financially-effective way of waste disposal.

Incineration is the dominant WtE technology globally and this trend is likely to continue owing to relatively low technology costs, market maturity, and high efficiency. Other thermal technologies such as gasification and pyrolysis are more efficient, score better in environmental impacts but have still high capital costs, and fit best countries with available capital and limited land resources, like the case of Japan.

According to WEC Report, another key barrier is that, "the WtE sector is very complex, fragmented in terms of policy and regulation and has a huge untapped potential. Both international and regional orchestrated efforts are necessary for the WtE market to be able to spread, benefitting thus the waste management and energy sectors."

As a conclusion, WtE development will need a deep consideration and implementation of polies and strategies, especially in developing countries to have a positive penetration in path in the next years.

4.4 WtE EXPERIENCES FROM LATIN AMERICA

4.4.1 Cases Studies

4.4.1.1 WtE Projects in Argentina

Estela Santalla

Departamento Ingeniería Química, Facultad de Ingeniería/UNICEN, Buenos Aires, Argentina

The state of art of the biodigestion in Argentina was identified through a survey on 62 operative plants (from a total of 105). This source revealed that 52.4% of them are in rural areas, mostly (85.3%) focused in wastewater treatment rather than in the production of biogas. Private sector concentrates 53.1% of the plants where 64.7% are large facilities (higher than 1000 m^3 capacity). More than a half of the public plants were installed for domestic wastewater treatment and for the valorization of the OFMSW while the rest are used for demonstrative and research goals; 58.3% are small size (lower than 100 m^3) and most of them have shown operative and management problems. The predominant biomass used as the source of organic matter for biodigestion are industrial waste from citric, beer, slaughterhouse, dairy, and yeast productions, followed in similar proportions by OFMSW and cattle waste such as manure from poultry, feed lot, swine, and milk cow wastewater. Regarding technological issues, more than 60% of the plants use continuous stirred-tank reactor (CSTR) and covered lagoons with 36.4% of foreign technology in the private sector (INTI, 2016).

The installed power of renewable energy in Argentina achieved in 2016 with approximately 246 MW, which represents 2% of the demand of the wholesale electricity market (MEM for its acronym in Spanish) (Cammesa, 2018). The current and historic values of renewable energy generated in the MEM shows that electricity from biomass projects increased 2.2% between 2011 and 2017 while energy from LFG decreased from the peak achieved in 2013 (Fig. 4.8).

A faster growth was observed in LFG projects during the last decade, due to the impulse triggered by the clean development mechanism (CDM), since the achieving of GHG certifications promoted the methane capture in larger and well-managed landfills. Thus, in Waste Handling and Disposal category there were developed 11 projects of extraction, collection, and flaring of LFG that registered GHG mitigation through the

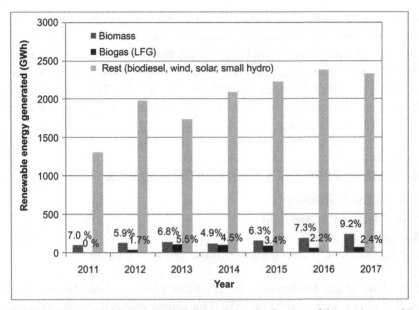

Fig. 4.8 WtE in Argentina. Percentages on each bar indicate the fraction of the total renewable energy by year (including biodiesel, wind, solar and small hydro). *(Source: Own elaboration based on Renewable Inform (Cammesa, 2018).)*

methane capture and destroying. In general, these projects were planned in two steps: a first one to achieve fully operational steady state of LFG capture system and a second one to advance in the electricity generation (Blanco et al., 2016). In addition, the GENREN Program promoted the development of some of these projects, despite its failure due to the distortion of the electric market by the high rate of subsidies.

Following are described the LFGTE projects operative or next to be:

- San Martín Central belongs to CEAMSE Environmental Complex Project Norte III A with an installed power of 7.1 MW, five internal combustion engines connected to the public grid.
- San Miguel Central belongs to CEAMSE Environmental Complex Project Norte III C with an installed power of 11.8 MW, six internal combustion engines connected to the public grid (see Fig. 4.9).
- Ensenada and González Catán Projects, each one of 5 MW power which concentrates the MSW that is generated in Ensenada, Berisso and La Plata, and La Matanza borough, respectively, the biggest one of the Metropolitan Area of Buenos Aires.
- Ricardone Project: this project is located at Santa Fe province and will generate electricity from LFG in a 1.2 MW plant that receives approximately 24,000 tons of MSW per month (from 2003) and will be operative in June 2018.

Fig. 4.9 San Miguel Central, CEAMSE Complex Norte IIIC. *(Source: http://www.ceamse.gov.ar.)*

The WTE projects adjudicated under the RenoAR Program are detailed in the following:

- Río Cuarto 1: the biogas is generated in a digester that consumes 20,000 tons maize silage per year (equivalent to 600 has) plus vinasse silage and swine wastewater to generate 2 MW capacity in Río Cuarto (Córdoba) with German technology. The project includes more than 30 partners from agricultural, services, and food sectors that are focused in the growth of the corn chain.
- San Pedro Verde Project: the project operates in Santa Fe province, generates biogas for electricity in a 1.42 MW plant, based on the manure collected from two dairies of more than 7000 milk cows. With an inversion of approximately U$S 6M, the projects require 160 direct jobs and 600 indirect for assembly (see Fig. 4.10).
- Río Cuarto 2 Project: a project of 1.2 MW under construction stage.
- Yanquetruz Project: project of 1.2 MW located at San Luis province, generates electricity from biogas produced in digesters that operate with swine wastewater and corn silage and provide electric energy to the public grid of the Integrated National System.
- Ricardone Project: described above.

The contracts for the supply of electricity (PPA) from biogas and LFG held between private companies, the Ministry of Energy and Mining and CAMMESA, for a term of 20 years, revolve around 160 dollars per MWh (except for Ricardone, which achieved 118 dollars per MWh).

Fig. 4.10 San Pedro Verde Plant at Chistophersen, Santa Fe province. *(Source: https://www.minem.gob. ar/energia-electrica/energias-renovables/prensa/26792/entro-en-operaciones-la-planta-de-biogas-san-pedro-verde.)*

4.4.1.2 WtE Projects in Brazil
Brazilian Landfills and Electric Power Generation
Vanessa Pecora Garcilasso

Research Group on Bioenergy, Institute of Energy and Environment, University of São Paulo, São Paulo, Brazil

In Brazil there are already some landfills with energy conversion from biogas, such as the recent power generation plant (Termoverde Plant—30 MW) at the Essencis landfill in Caieiras/SP, which was inaugurated in September 2016, which is the largest thermoelectric fueled by biogas from urban solid waste in Brazil.

In São Paulo, another two landfills generate energy from biogas: Bandeirantes Landfill closed in 2007, and São João Landfill. The initial power of the Bandeirantes Landfill biogas plant was 22 MW and was inaugurated in 2004. The plant at the São João Landfill was 20 MW and was inaugurated in early 2008. The landfills have already been closed but the biogas conversion is still in operation with the remaining organic matter existing in the landfills.

The first thermoelectric power plant in Northeast region (and the third of this type in Brazil), Termoverde Salvador, is installed in the Metropolitan Sanitary Landfill of Salvador and began its operations in March 2011. Termoverde Salvador has an installed capacity of 20 MWe from the landfill biogas (Solvi, 2010).

Estre, a Brazilian company that develops environmental solutions for all stages of waste management, has launched at Guatapara landfill its first biogas power generation plant in 2014, the so-called 4.2 MW-Bioenergy Plant. The investment is supported by Investe São

Paulo, agency from the State Government responsible for attracting investments linked to the Secretariat of Economic Development, Science, Technology, and Innovation.

In 2015, the first power production plant using landfill biogas at Rio Grande do Sul was launched, at Minas do Leão landfill, from Companhia Riograndense de Valorização de Resíduos (CRVR). Biotermica Energia Plant has an installed capacity of 8.55 MW and will reach 15 MW, when in full operation (Osepeense, 2015).

Biomethane Production Plants

Vanessa Pecora Garcilasso

Research Group on Bioenergy, Institute of Energy and Environment, University of São Paulo, São Paulo, Brazil

Currently there are some relevant projects under development, regarding biomethane use in Brazil, in different regions.

In the state of Rio de Janeiro there are two landfills producing biomethane from urban solid waste.

The Dois Arcos Plant, from Ecometano, located in São Pedro D'Aldeia, in operation since 2015, produces biomethane from urban solid waste from eight municipalities in the Lagos Region. The unit has an installed capacity to produce 15,000 m^3/day of biomethane. At the outset, biomethane will be supplied in cylinders, as compressed natural gas, to industrial customers. In the future, the plant may be connected to the distribution network of the State's piped gas utilities.

Another project concerns the replacement of part of the natural gas consumed in the Duque de Caxias (Reduc)/RJ Refinery of Petrobrás, by the purified biogas from the Gramacho Landfill.

Closed in 2012, this landfill no longer receives garbage, but continues to produce gases. The biogas produced in the landfill is purified until it reaches the quality standard required by the Petrobrás technical specifications and then is disposed of in an exclusive 6-km pipeline to Reduc (Petrobras, 2014).

With the use of biomethane to generate energy at Reduc, the estimate is that in the next 17 years (from the start of the system's operation in 2014), approximately 6 million tons of carbon dioxide will no longer be emitted into the atmosphere. The project also includes the transfer of biogas purification technology and has a CDM project that consists of the capture of biogas from the Gramacho landfill, purification, and injection into Reduc's exclusive natural gas distribution network, thus replacing part of its consumption of natural gas. The excess biogas produced will be flared in flare (Petrobras, 2014; DCP/UNFCCC, 2012).

In 2018, Gas Natural Renovavel—GNR/Renewable Natural Gas, in Caucaia landfill[8] in Fortaleza, Ceara state, has launched the largest biomethane plant in Brazil, producing up to 85,000 Nm^3 of biomethane per day sold to CEGAS, the state natural gas utility.

---

[8] https://www.opovo.com.br/jornal/economia/2018/04/usina-de-biogas-entra-em-operacao-no-proximo-dia-16.html.

Brazil—WtE Incineration Plant

Fábio Rubens Soares and Suani Teixeira Coelho

Research Group on Bioenergy, Institute of Energy and Environment, University of São Paulo, São Paulo, Brazil

Incineration plants in Brazil are in their early stages. Despite some previous initiatives, there is only one WtE incineration plant under implementation—the so-called MSW Energy Recovery Unit (ERU) in Barueri—São Paulo State.

This ERU project is designed to perform the thermal treatment (burning) of the MSW (household waste), collected in the municipalities of Barueri, Carapicuíba, and Santana do Parnaíba (municipalities of the Metropolitan Region of the State of São Paulo). MSW will be fed in a rate of up to 825 ton/day, generating an installed power of 17 MW. The ERU will run a Rankine Cycle that will use the burning of MSW for steam generation, which will feed a turbo-generator for electric power production. The expected lifetime of the project is 30 years and the ERU's operating forecast is 8000 h/year.

Barueri-ERU is designed to work continuously at the nominal power, being able to meet variations of load and heating value of the MSW collected. Trucks will transport MSW from the local municipalities to be received in the ERU.

The electricity to be generated in the Barueri-ERU will be connected to a local sub-station in 88/138 kV, located adjacent to the plant. The electricity will be commercialized through a long-term contract in a free market.

The ERU licensing process was performed with CETESB, the State of Sao Paulo Environmental Agency, following the municipal, state, and federal licensing rules, in particular the pollutant emissions limits established. The Environmental Secretariat of the State of Sao Paulo has adopted Resolution SMA 79, in 2009, following the same limits of European Union for dioxins and furans.

There were not huge difficulties for the formal licensing, despite the usual concerns of the civil society (as worldwide, due to the lack of adequate information), since there is the adequate local legislation. In fact, the main problem faced was to obtain the whole investment needed for the plant. According to personal information, in 2017, finally entrepreneurs were able to get the funds needed for the plant to start the construction.

Case Study Brazil—WtE Gasification Plant

Luciano Infiesta* and Javier Farago Escobar[†]

*Carbogas Industries, São Paulo, Brazil
[†]Research Group on Bioenergy, Institute of Energy and Environment, University of São Paulo, São Paulo, Brazil

Electricity generation from RDF through gasification process has started in a pilot plant in semi-industrial scale for the development of a circulating fluidized-bed gasification process. A prototype plant designed by the company *Carbogas Energia* is located in Mauá municipality, in São Paulo state. In this plant, several field tests were developed with the monitoring of the environmental company of São Paulo—CETESB, *researchers from the University of São Paulo* and Societé Générale of Surveillance (SGS) to control the atmospheric emissions. More recently, also Furnas Centrais Elétrica Ltda supervised these works, in the context of the new plant under construction in Boa Esperança municipality, Minas Gerais State. The results from the tests were considered adequate and Carbogas was authorized to proceed the assembling of real size unities such as the one in Boa Esperança.

The parameters to be followed are the ones established by the (Federal) Resolution CONAMA no. 316 and (Sao Paulo State) Resolution SMA-79, which correspond to the regulatory legislation used for the legal validation of thermal processes in Brazil. The obtained data compared to the Resolution CONAMA no. 316 presented satisfactory results in all the requirements; for example, the amount of dioxins and furans (D&F) in the three samples studied were up to 50 times lower than the allowed standards. The same tests were performed and compared with the resolution SMA 79, which follows the European patterns of atmospheric emissions, which are the most strict standards worldwide. Following the same comparatives, D&F emissions reached figures up to 12 times lower than those allowed in the resolution. Based on the obtained results, the fluidized-bed gasification process appears to be quite promising under environmental aspects. This is also a quite important social issue since local society in Brazil is quite often concerned about D&F emissions and the health impacts they can produce if not adequate eliminated.

In 2017, based on these results, state-owned Furnas electric utility (Furnas Centrais Elétricas) started the construction of the 1 MWe-plant under a research and development project in Boa Esperança municipality, Minas Gerais State, the so-called *Usina de geração termoquímica de energia/UGTE* (Thermochemical power plant). This project aims to be pioneer in industrial scale, giving fundamental subsides to the development and dissemination of its technology. It has been designed to handle 60 ton/day of MSW corresponding to 1.25 MW of installed power. The processed RDF will follow to a packer press where it will be compacted and packed with an average humidity of 12%. The packs will be approximately with a volume of $1 \, m^3$ and weighting averagely 800 kg. The construction will allow storage of approximately 3200 packs (see Fig. 4.11). All the liquid effluents from the waste management process and RDF transformation will be treated inside the facility area and transformed into reuse water.

Inert products and ashes (8.8%) are extracted through the bottom of the fluidized bed. The syngas generated in the gasification process has a low heating value in the range of $1160 \, kcal/Nm^3$ up to $1260 \, kcal/Nm^3$ with a volumetric flow rate of $2736 \, Nm^3/h$ and mass flow rate of 3218 kg/h. The particulate dragged, and the chlorine products are eliminated through the addition of calcium hydroxide in the gas cleaning system, through a ventury-type closed circuit and scrubbers.

The clean gas is cooled and sent to a recovery boiler, to be burned producing 5.5 ton/h of steam at 42 bar, 420°C. This steam is fed into a condensing steam turbine, so the global energy balance shows process is self-sustainable.

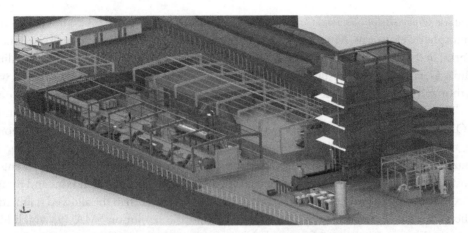

Fig. 4.11 Gasification in fluidized-bed facility layout. *(Source: Carbogas Co.)*

This fluidized-bed gasification technology was developed for modules from 50 to 300 ton/day of MSW. The tests result also show that ideal results are for modules capable of at least 300 ton/day of urban solid residue. In these conditions, it is possible to reach an efficient processing capacity, for municipalities with 400 thousand inhabitants. Notwithstanding this, smaller modules have also presented interesting economic and experimental results to support smaller towns (up to 50 thousand inhabitants). In places with a demand higher than 300 ton/day, more than one module can be assembled, in a way to obtain a higher capacity factor with better operational performance in each module. This practice can contribute to maximize the results and minimize cost with the facility.

Taking this experience into account, it may be concluded that the thermochemical gasification using urban solid residues is a remarkable candidate as a final destination to this environmental liability and also an attractive business plan to power generation, offering economic, social, and environmental beneficial effects. Therefore, it deserves the proper attention and subsidies to keep on development and improve.

4.4.1.3 WtE Projects in Ecuador

Rafael Soria and Laura Salgado

Departamento de Ingeniería Mecánica, Escuela Politécnica Nacional, Ladrón de Guevara, Quito, Ecuador

Methane Capture in Landfills and Its Use for Energy Purposes

Only two GADs are capturing its landfill methane for power generation. The first one is "El Inga" project in Quito, which receives an average of 2000 tons of MSW/day. Currently El Inga has an installed power capacity of 5 MWe. In a first phase, from 2016, with an installed capacity of 2 MWe it provided an average of 15.6 GWh/year of electricity to the National Interconnected Power System (SNI). In a second phase, from March 2017, 3 MW was added to the project to provide additional 23.4 GWh/year to the SNI. This plant was built by Gasgreen SA company and by the local company EMASEO. The biogas collected is conducted to a drying and desulfurization system, an then it fuels internal combustion engines of 1 MWe each, coupled to power generators CO_{2e} emission tons per year.

The second project is "Pichacay," which is located in Cuenca. Pichacay receives an average of 490 tons of MSW/day and has an installed power capacity of 1 MWe, producing electricity for the SNI since 2015 (EMAC EP, 2015). This project seeks to expand its installed power capacity to a total of 2 MWe in the mid-term. This plant was built by the Dutch company BGP Engineers and by the local company EMAC. This project has 30 wells to collect biogas and conduct it to a drying and desulfurization system. This biogas, with an average methane content of 55%, fuels internal combustion engines of 1 MWe each, coupled to power generators (EMAC EP, 2013a,b). This project supplies an average of 7013 MWh/year.

Both projects, El Inga and Pichacay, benefit from the regulations CONELEC 004/11, valid for projects registered until December 31, 2012, which aimed to support power generation projects with nonconventional renewable energies by guaranteeing a feed-in-tariff of 11.05 cent USD/kWh. This regulation is no longer valid from 2013.

Controlled Biogas Firing at Flares

Additionally, five GADs collect the biogas generated at their landfills for controlled methane firing at flares. This is the case of landfills located in the GADs of Santo Domingo de los Tsachilas, Ibarra, Otavalo, Ambato, and Portoviejo. The implementation of these active biogas capture systems at existing landfills is possible due to the "Sectorial Mechanism of Mitigation (MSM) in the Solid Residues Sector in Ecuador." The Latin American Investment Facility (LAIF) of the European Union supports the Assistance Program with funds provided by KfW. Regional implementing partner is the Development Bank of Latin America (CAF). For Ecuador there is 5 million Euro nonrefundable to finance technical assistance (1 million Euro) and to fund economic incentives (4 million Euro), using the financial mechanism of performance-based climate finance (PBC), offered to pilot projects that mitigate beyond the baseline scenario (CAF, 2015a,b). With this financial mechanism, the payment is done ex post, after the measure, verification, and report (MRV) of the amount of methane reduction of each project. In this context, the LAIF-PBC financed the design of the MSM in Ecuador, identified pilot activities, evaluated the mitigation potential, and costs of the mitigation activities. Depending on the project, the minimum marginal abatement cost (MAC) was 4.20 USD/tCO$_2$e, while the maximum was 20.00 USD/tCO$_2$e for the operational years 2018–21 (Wade-Murphy, 2018). The incentive depends on the project, but an average is around 7 USD/tCO$_2$e. This incentive considers the MAC plus an additional incentive to foster the optimal operation of each project (CAF, 2015a,b). This program has been ongoing in Ecuador since 2014 and with the implementation of these five projects it is expected a reduction of 500,000 tCH$_4$/year. Additionally, CAF will provide a reimbursable credit to the Government of Ecuador, through the Bank of the Ecuadorian State (BDE), to which the GADs may access to finance the infrastructure construction. The first phase of the technical design studies also analyzed the resources that would be needed to utilize the recovered LFG for electricity generation. The range of estimated minimum sales price for electricity required to permit the mitigation activities to be profitable under the energy use scenario (same CH_4 mitigation potential, adding the investment, and O&M costs related to electricity generation) would be between 7.96 and 11.85 cent USD/kWh (Wade-Murphy, 2018).

The "Mundo Verde" commonwealth, which is integrated by 20 GADs from Guayas, Los Ríos, and Bolivar provinces also received economic support to close open dumps and start building landfills. The landfill of Guayaquil, the largest city of Ecuador, has good opportunities to implement controlled methane fire at flares, or even for power generation. The MAE is engaging Guayaquil authorities to go ahead with this initiative. Loja, Cuenca, Quito and Guayaquil, and Santa Cruz of Galapagos are part of the "Footprint of Cities" project, financed by CAF and the French Development Agency (AFD) (CAF, 2018). In the scope of this project, one of the prioritized projects in Loja will be the capture of biogas from its existing landfill for an energy use (CAF, 2017). So far there is no date to start this project.

In 2013, UNFCCC financed the second Technology Needs Assessment (TNA) for Ecuador—Waste to Energy. This study prioritized three technologies for Ecuador: (a) pellets production and its co-firing in cement industry furnaces.; (b) biogas active capture in existing landfills and its energy use; (c) mechanical biological treatment (MBT) with anaerobic digestion and biogas production for energy uses (ENT/MAE/URC/GEF, 2013a,b). Nevertheless, these technologies have not been implemented so far at large scale.

Finally, in 2013 a 10.7 MW WtE project based on pyrolysis and gasification was proposed to use the MSW of Chone. Nevertheless, this project faced financing problems and never happened.

Energy Recovery From Agricultural Waste

The Ministry of Environment has reviewed the most outstanding initiatives and projects implemented about WTE with the agricultural waste. The initiatives and projects studies are private or come from the private sector. One of the technologies applied deal with the operation of a biorefinery, for research purpose, that produces bioethanol from tagua and sugar cane bagasse. Another investigation initiative is implemented in the municipality of La Concordia in 2009, where biofuel is produced through gasification that allows the conversion of organic waste from agricultural industry to generate syngas, and it is used for electricity production. In the same location a pilot plant of pyrolysis was implemented too, however this plant only uses palm waste and the main outcome is charcoal. From the private sector, sugar refineries have played an important role, since 2004 the usage of sugar cane bagasse for cogeneration has resulted in the production of heat and electricity. Ecuador has an installed power capacity of 136.4 MW based on cogeneration plants at three sugar mills (San Carlos, Ecoelectric, Ecudos) (MEER, 2017). Other companies in the food and beverage sector produce small quantities of biogas and biodiesel from agro-industrial residues (Ministerio del Ambiente, 2014).

4.4.1.4 WtE Projects in Mexico

Gustavo Solórzano

Asociación Mexicana de Ingeniería, Ciencia y Gestión Ambiental, A.C. (AMICA), Mexico City, Mexico

LFG to Energy

Final disposal of MSW in landfills has been improving in recent decades in Mexico, as in most countries in the LAC region. In 1997, final disposal in sanitary landfills represented only 40.7% of the total generated MSW, while by 2013 this percentage had increased to 74.5%. This is equivalent to an 82.7% increase in waste destined for sanitary landfills over a 16-year period (SEMARNAT, 2016). This situation has enabled Mexico to develop several landfill-gas-to-energy projects, encouraged by the CDM established within the UNFCCC, which considered the combustion of methane and/or energy production from the LFG in sanitary landfills as a viable option for reducing GHG emissions and thus displacing fossil fuels. Although an important number of landfills in Mexico (shown in Table 4.3) have been granted registration in the CDM, only a few of them are currently in operation (GIZ-EnRes, 2019).

Given the potential advantages of this kind of projects, and the volume of MSW generated every day but also partially accumulated in Mexico's almost 2500 municipalities, it seems that eight energy-producing landfill projects is a small number. However, it is expected that the recent energy reform and an improving regulatory framework will stimulate these and other waste to energy projects in Mexico.

Coprocessing MSW

For many years, cement kilns in Mexico have been coprocessing wastes such as spent oil/solvents, paint, tires, and, at a lesser degree, MSW. The Mexican cement industry reports coprocessing levels of 12% in terms of thermal energy replacement (GIZ-EnRes, 2016).

Among others, cement producer CEMEX started a project with Mexico City in 2012, in which it receives a mix from the inorganic fraction in MSW, with an average composition of 32% plastics; 50% paper and cardboard; 10% textiles; and 8% wood.

Table 4.3 Landfill Gas Energy Recovery Facilities in Mexico

Location	Start Up (Year)	Installed Capacity (MW)
Salinas Victoria, Nuevo León	2003	16.96
Ciudad Juárez, Chihuahua	2011	6.4
Aguascalientes, Aguascalientes	2011	2.615
Durango, Durango	2014	1.6
Atizapán de Zaragoza, State of Mexico	2014	0.6
Cuautla, Morelos	2014	1.065
Saltillo, Coahuila	2014	2.122
Querétaro, Querétaro	2012	2.746
Total		34.108

CEMEX prepares and transports the waste to two cement plants and feeds the waste into its own cement kilns. A total of 84,000 tons of inorganic MSW were processed in eight CEMEX plants in the country in 2013 (GIZ-EnRes, 2019).

According to GIZ-EnRes (2016), there is a potential replacement for thermal energy in the cement industry up to 30% in Mexico, which means using 3.1 million tons per year of MSW. This would be equivalent to 8.1% of the total amount generated in the country, which has the potential to mitigate the emission of 3.2 Mt of CO_2e per year. Nevertheless, a lack of source separation programs for MSW, pretreatment/transportation costs, and an inadequate regulatory framework are important challenges to this WtE alternative (GIZ-EnRes, 2016).

Anaerobic Digestion

A significant number of anaerobic digestors (currently around 700) have been operating in Mexico with different kinds of animal excreta for several years. But the digestion of the organic fraction in MSW (OFMSW) as an option to treat this fraction in MSW has only recently been explored, as is the case in other countries in LAC region.

The first anaerobic digester for OFMSW has been built in Atlacomulco, Mexico, with a capacity to process 30 ton/day. The digester was built with a US$ 2.8 million grant from the Federal Government; the municipality provided the land. The digestor contemplates a wet process and according to the designer it will produce up to 2400 m^3/day of methane with an energy yield of up to 1,752,091 kWh/year (GIZ-EnRes, 2019).

Incineration

Mexico City (CDMX) initiated the process to build and operate an incineration WtE plant. In December 2016, CDMX published the "specifications for the award of the contract for … the design, construction, commissioning, operation and maintenance of a plant for recovery of the calorific value of municipal solid waste … for the generation and delivery … of electricity … for up to 965,000 MWH a year" (CDMX, 2017). The bidding document was expressly written this way due to the city government's aim to stop paying for the energy needed to run its underground transportation system, transporting almost 6 million passengers every day. The corresponding contract, valid for 33 years, was granted to the multinational French firm Veolia, who would operate the plant to burn nearly 4500 ton/day out of the 13,000 tons of MSW generated daily, of which 8500 tons are currently being exported to landfills located outside the city (Veolia, 2017). As such, the plant was designed to provide the energy requirements to run the underground transportation system, an action that according to CDMX and Veolia would mitigate the equivalent of 700,000 tons of CO_2 emissions. Unexpectedly, the new Mexico City administration which took over in December 2019 decided not to continue with this project.

4.4.1.5 Particular Case of Study From LA: Energy Recovery of Waste Tires in Latin America: Combustion in Cement Kilns and Thermochemical Conversion by Pyrolysis

Juan Daniel Martínez[9]

Grupo de Investigaciones Ambientales (GIA), Universidad Pontificia Bolivariana (UPB), Medellín, Colombia

The amount of waste tires (WTs) generated in developing countries increases as the automotive sector grows. The worldwide tires demand shows a growth around 4.3% per year, having reached 2.9 billion units in 2017 (Machin et al., 2017). WTs are a very special waste highly resistant to natural biodegradation or photochemical decomposition. In addition, their complex toroidal shape (around 75% of the volume is empty space) makes expensive their logistic and hence, very complex their final disposal.

Dumping WTs in inappropriate places leads to environmental impacts, including rainwater accumulation (creating favorable conditions for propagation of disease-carrying pests), leaching, and open-air burning. When ignited, WTs are very hard to extinguish, and the uncontrolled fire can even last several days. Many pollutants emitted from WTs fires are toxic, carcinogenic, and/or mutagenic. Significant enrichments in ambient concentrations of CO, CO_2, SO_2, particle number (PN), fine particulate (PM2.5) mass, elemental carbon, and polycyclic aromatic hydrocarbons (PAHs) have been reported (Downard et al., 2015). Besides the release of hazardous smoke, tar-condensed compounds can affect the soil and the groundwater. In countries without practical and sustainable WTs management programs, the frequency of these fires is suggested to be much greater (Shakya et al., 2008). The most recent WT fire in Latin America (LA) was in November 2014. Around 600,000 waste tires burned during several days in Bogotá (Colombia) leading to one of the greatest environmental tragedies in the history of the city (Park et al., 2018).

The progress made in recent years in the management of these wastes, mainly in developed countries, has meant that WTs are starting to be perceived as a potential source of valuable raw material (Sienkiewicz et al., 2012). Some of the WTs generated in LA are retreaded or subjected to shredding operations for specific applications including material and energy recovery. Material recovery comprises rubber chips production for roofing materials, playground covers, sport flooring, asphalt, floor mats, carpet padding, and plastic products among others. Energy recovery implies the WT use in energy demanding facilities for combustion, gasification, and pyrolysis processes. At industrial scale, WT combustion comprises the use of this waste as counterpart of

[9] The author is in debt with Natalia Cardona (UPB, Colombia), Mauricio Giraldo (Cementos Argos, Colombia), Lina Rojas (FICEM, Colombia), Germán Aguilar (ReacecolGreen, Colombia), José Betancourt (360° Process Engineering Solutions, México), Emanuel Bertalot (ReNFUPA, Argentina), Diego Rodríguez (Gemini, Belgium), Arturo Rock (Kona Fuel, Chile), Patricio Vascónes (Ecuaneumáticos, Ecuador), Christian Coronado (UNIFEI, Brazil), Simón Augustower (Ecocombustion, Panama), Oberto Marini (Komalt SAS, Dominic Republic). This work would not have been possible without the valuable help of these people.

different fuels in a variety of industrial applications such as electric utility boilers, pulp and paper mills, and cement kilns. This practice, known as coprocessing, takes advantage of the notable heating value of the WT (up to 40 MJ/kg) and serves as strategy not only to curb CO_2 emissions, but also to develop energy-saving techniques. On the other hand, WT gasification at commercial scale is rather scarce and seems to be only at both laboratory and pilot scale (Oboirien and North, 2017). Finally, WT pyrolysis is gaining each time major maturity. There exist commercial and demonstration plants running worldwide aimed aimed at the production of hydrocarbon liquids and carbonaceous solids for different applications.

Waste Tire Generation

It is estimated that more than 1000 million of WTs are generated each year in the world (17 M tons). One tire per person is discarded each year in developed countries (Martínez et al., 2013a). The increasing need for efficient waste management has made that 92% of WTs in EU be recovered by different practices such energy, material, and retreading (ETRMA, 2017); while in the United States, these practices are around 89% (RMA, 2015). However, the final destination of WTs is still a big challenge in most of developing countries. Most of these wastes are stockpiled in nondesignated sites causing land, water, and air pollution, putting at risk the entire community.

For LA, Brazil and Mexico show the highest WT generation, around 585,000 and 310,000 ton/year, respectively. Argentina and Chile show similar tends, about 140,000 ton/year, while Colombia is around 120,000 ton/year. The WT generation in other LA countries is not discussed due to the lack of reliable information.

Regulations

Waste management in LA has historically been hampered by weak infrastructure: inadequate collection services and limited landfill capacity (Rouse, 2005). Brazil and Mexico launched the first policies for prevention and integral management of wastes. These actions have promoted waste management legislations in other LA countries, including the management of hazardous wastes and the extended producer responsibility (EPR) for end-of-life products, among others. Even though there are different legislations to support reduction, reuse, recycling, and disposal strategies, many times is not sufficient to guarantee the proper waste disposal in many LA countries.

Brazil, Mexico, Chile, and Colombia have implemented EPR systems for different end-of-life products including WTs. Brazil and Mexico are under the frame of the shared EPR model (extended and shared producer responsibility), while in Chile and Colombia, the responsibility of local governments plays a minimal role. Without any economic instruments, Colombia has adopted and implemented the EPR program for WT in 2010 from the resolution 1457 (2010). It is expected that Colombia accelerates the WTs valorization from both material and energy recovery, given the resolution 1326 (2017) aimed the selective collection and environmental management of WTs (Park et al., 2018). The EPR system in Chile has been promulgated recently from the law 20920 (2016), and it will be applied progressively as the industrial capacity is developed to meet with the increasing valorization goals. On the other hand, Argentina has lately launched a law project for management and valorization of the WT (1723/11), although definitions and guidelines for sustainable tire management have been established in the resolution 523 (2013).

Waste Tire Combustion in Cement Kilns (Coprocessing)

The energy recovery of WTs represents an important share in EU and the United States. Currently, around 49% of the WTs generated in the United States are destined for cement production, pulp and paper mills, electric utilities, and industrial boilers (RMA, 2015), while the in EU these practices accounts for 28% (ETRMA, 2017). Among these energy recovery practices, the cement production is the most frequent. Although some of these industries can burn the entire WT, it is also possible the use of rubber chunks, in order to fit in most combustion units. These rubber pieces must be uniform and flowable, favoring their handling for transportation, and are commonly known as TDF (tire-derived fuel). TDF has demonstrated compatible characteristics to be used in existing combustion units and important guidelines regarding properties and characteristics to end-users are found in an ASTM standard D-6700-01 (ASTM, 2013).

The coprocessing of WTs in cement plants has been used since 1970s in the United States, EU, and Japan. The long fuel residence time and the very high temperatures inside the clinker kiln favor the WT combustion. In addition, the WT ashes are incorporated into the clinker matrix, and thus no waste is generated. Moreover, WT coprocessing is recognized to aid in curbing the use of fossil fuels mainly given the natural rubber (biomass) contained into the WT. The cement industry demands large amounts of energy and for this reason, this sector is a major contributor of greenhouse gases (GHG) emissions. One ton of cement releases between 500 and 900 kg CO_2 to the atmosphere, making this sector the responsible between 5% and 7% of the total global CO_2 emissions (Benhelal et al., 2013).

Since 1990s, Brazil and Mexico have been coprocessing not only WT but also residual biomass, sewage, and industrial sludges, oils/solvents used, among others. Around 35 cement plants in each of these two countries use WTs as supplementary fuel. The WTs use in cement production is the most common disposal procedure in Brazil. Around 46% of total WTs generated in 2017 was used for combustion in cement plants

(IBAMA, 2018). In Mexico, between 10% and 15% of WTs are used for coprocessing in cement kilns (CEMBUREAU, 2018). Colombia is also coprocessing WT in cement plants and currently there are two plants under operation in a continuous way. Finally, at least one cement plant in Argentina, Chile, Costa Rica, Dominican Republic, Guatemala, Honduras, and Uruguay are using WTs in cement kilns, with a tendency to increase in near future.

Waste Tire Pyrolysis

Unlike combustion, pyrolysis is a thermochemical treatment in a nonoxidizing atmosphere. When properly applied to WTs, pyrolysis enables obtaining liquid and gaseous hydrocarbons of high energy density, and a solid carbonaceous fraction (Martínez et al., 2013a). Although this process is endothermic, the total energy balance shows a net positive energy process. This is because the gaseous fraction released in WT pyrolysis has high energy density and is capable not only of satisfying the energy demanded by the process but also of producing electricity (Martínez et al., 2013b).

The rotary kiln has been the technology with the greatest penetration in LA. In this reactor, WTs are fed as coarse shreds or entire and can be introduced to the kiln mechanically or manually. This kind of reactors commonly works in batch mode with residence times up to 9 h, and cooling times up to 12 h. The temperature inside the reactor generally does not exceed 450°C. The remaining wire steel is between 10 and 15 wt%. The liquid yield is usually between 20 and 35 wt% although is possible achieved 45 wt% after a proper maintenance work of reactor, pipes, and condenser. The solid yield also varies depending on several factors as the pyrolysis conditions, the arrangement of the WTs inside the reactor and even the WT type. Typical values are between 25 and 35 wt%. Even though rotary kiln reactors are the most used technology for WT pyrolysis in LA, it faces several challenges in order to achieve a successful process. For instance, some of these plants lack of control devices with proper automation and safety systems. Likewise, there are no gas evacuations systems in case of unscheduled stops. High-quality materials for reactor, pipes and condenser, and ATEX equipment are commonly missing.

The feasibility of pyrolysis (both technical and economical) depends on multiple factors. Among others, it is worth to highlight a reliable technology proven in the long-term and a proper operational condition able to achieve a balance between a high liquid yield and good physical-chemical properties of the solid fraction. In most cases, the liquid fraction has a well-established market since it can be used as direct fuel, petroleum refinery feedstock, intermediate fuel oil (IFO), and even as a heavy crude diluent. Some characteristics and properties as a fuel as shown elsewhere (Martínez et al., 2013c). Although the solid fraction can be used as fuel for bricks, ceramics, blocks, and even manufacture of cement, it often represents the big challenge to be incorporated in a real and stable market. It is commonly find that some technological providers offer this product as carbon black (CB) but it is really not. The resulting solid fraction from WT pyrolysis is indeed a complex mixture where all the CBs and fillers used in the tire manufacture are trapped

(Martínez et al., 2019). For this reason, the solid fraction must be produced with minimum volatile compounds as well as be submitted to refining, milling, and palletization processes in order to achieve a suitable material able to compete with commercial CBs.

The pyrolysis of WTs in LA does not show the same picture compared to that of combustion in cement plants. The country with more industrial pyrolysis plants is Mexico. There are between 15 and 20 erected plants, although at least four pyrolysis plants are currently under operation meeting the environmental and industrial safety regulations. The nominal throughputs of these plants are around 20 ton/day and operate in both semi-continuous and batch-mode. Brazil seems to be the country with the highest pyrolysis processability in LA. More than 13,000 ton of WTs were pyrolyzed in 2017 (IBAMA, 2018). Colombia, Panamá, Dominic Republic, Ecuador, and Chile also have industrial pyrolysis plants with nominal capacities between 10 and 20 ton/day. At least one pyrolysis industrial plant operates in each of these countries by using the rotary kiln reactor.

Final Remarks

Large part of the WTs generated in developing countries are still disposed in landfills or stored in illegal places endangering environment and society. Collection of WTs still faces many challenges given the large expenditures and the operational, environmental, and social problems involved (Costa-Salas et al., 2017). It is also worth to highlight the lack of joint action between the main stakeholders: final consumers, the government, and private companies (Fagundes et al., 2017).

Coprocessing is the most common energy recovery practice for WT in LA. Around 35 cements plants in both Brazil and Mexico are currently operating. Conversely, there are few industrial WTs pyrolysis experiences. Mexico seems to have more industrial pyrolysis plants than any other country in LA. The adoption and intensification of these energy recovery practices in LA faces legislative, technological, economic, environmental, and market barriers, and hence; regulations/policies for sustainable WT management and valorization are strongly needed.

CHAPTER FIVE

WtE Best Practices and Perspectives in Asia

Shyamala K. Mani
National Institute of Urban Affairs (NIUA), India Habitat Centre, New Delhi, India

5.1 MSW MANAGEMENT AND POLICIES IN ASIAN COUNTRIES

Roshni Mary Sebastian, Dinesh Kumar, and Babu J. Alappat
Indian Institute of Technology, New Delhi, India

Sustainable solid waste management is a grave issue in developed and developing Asian countries. While the economic growth indicators reflect development in the region, a corresponding increase in municipal solid waste (MSW) generation was observed, simultaneously. South Asia alone generates about 70 million tonnes of MSW per year, with an average of 0.45 kg/cap/day. About 26% of the world's MSW is generated collectively from South Asia, East Asia, and the Pacific region (Hoornweg and Bhada-Tata, 2012a, b). Variations in the consumption patterns due to economic progress and lifestyle changes have marginally reduced the putrescible fraction and increased the combustible fractions. With the adverse environmental impacts of unscientific disposal of MSW gaining attention, employment of material and energy recovery operations can steer the waste management scheme to sustainability. Consequently, the quantum of MSW disposed in landfills reduced from nearly 90% in countries like China (Karak et al., 2012) and India (Sharholy et al., 2008) to 68% (National Bureau of Statistics of China, 2015) and 75%, respectively (Yadav and Samadder, 2017).

Integrating energy recovery techniques, precisely thermal treatment techniques into the comprehensive waste management scheme can aid in quick disposal of MSW with energy recovery. Besides helping meet the rising energy demand to some extent, this can serve as a key to the circular economy on a long-term basis. Waste incineration is often realized as the realistic treatment technique, with diverse growth trends in various countries. Japan started incineration for garbage disposal in 1960 (Ministry of the Environment, 2012) and incinerates nearly 80% of the generated MSW, as of 2013 (Tabata and Tsai, 2016). Incineration for waste disposal in China began in 1980, with 30% of the generated MSW being incinerated for disposal as of 2013 (Zhang et al., 2015). The first incineration plant in India was set up in 1987. However, despite the state-of-the-art technology, the plant was shut down after a few weeks of operation, due to the poor thermal characteristic of the waste feed (Talyan et al., 2008). Similar was the case with the incinerators in the tourist islands of Malaysia, where the thermal characteristics of MSW were inadequate to practice incineration (Sharifah et al., 2008). The origin and growth of the waste to energy (WtE) sector has been quite dynamic in different countries, owing to the quantum and characteristics of the MSW generated as well as the environmental policies and legislation formulated in the respective economies. This chapter gives an overview of the status of waste incineration and Biomethanation for MSW disposal with energy recovery in various countries in Asia. A brief feasibility assessment of waste incineration with respect to the thermal characteristics of MSW has also been carried out using conventional methods. The chapter also gives a concise account of the techno-economic aspects of waste incineration and biomethanation.

5.1.1 Challenges in Waste Management

1. The absence of the formalized waste diversion sector.
2. In Asia, the influence of the informal sector often far exceeds that of the formal sector.
3. In Pune, India, the informal sector recovers 22% of its waste.
4. A much better approach is to integrate waste pickers into the waste sector.

A major characteristic of the informal sector is its invisibility. Thus, raising the integration of the sector requires raising awareness of the political decision-makers of the contribution of the sector as an appropriate waste management system.

5. Lack of enforcement of waste management legislation and lack of citizen's participation.
6. Multiple government departments are assigned overlapping responsibilities in waste management activities.
7. A mechanism to overcome is to have an interdepartmental committee or task force that oversees and coordinates work is envisaged. Fig. 5.1 gives a possible national and city level action plan toward achieving sustainable waste management.

5.2 ENERGY ACCESS IN ASIA: CURRENT SITUATION AND DIFFICULTIES TO INCREASE ACCESS

Shyamala K. Mani
National Institute of Urban Affairs (NIUA), India Habitat Centre, New Delhi, India

According to a market intelligence report, titled "Asia-Pacific WtE Market—Analysis and Forecast (2018–2023)," the Asia-Pacific WtE market is expected to reach USD13.66 billion by 2023, rising at a CAGR of 15.5% from 2018 to 2023. An increasing amount of residential and industrial wastes has led to the growth in China and India besides government support for sustainable energy practices in the region. For residential,

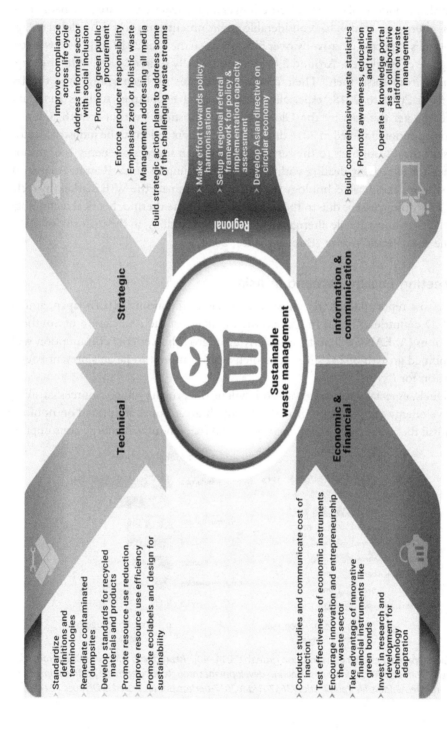

Fig. 5.1 Possible actions that may be used to draw a national or a city level plan toward sustainable waste management. (*Courtesy: P. Agamuthu, University of Malaya, Malaysia (Personal Communication).*)

commercial, and industrial sectors, electric power and transportation are core necessities. However, energy generation has considerable environmental impacts. The global energy consumption is expected to grow by over 50% between the year 2010 and 2040, with the increasing use of fossil fuels. Around 29 countries globally source >90% of their energy requirement from fossil fuels. Their consumption rate has been rising over the last 10 years from 2010 to 2018, especially in developing economies such as India and Singapore. As per the data by the Global Energy Statistical Yearbook 2018, Chinese energy consumption doubled in 2016 due to the rising demand from the industrial sector.

The shift from fossil fuels to the renewable and clean sources of energy is of utmost importance for effectively dealing with climate. According to Sonal Rawat, Analyst, in 2018, the thermochemical technology is expected to dominate the WtE market (by technology). This dominance is due to the increasing use of the thermochemical technology in WtE plants, as a sustainable alternative to unsanitary landfills, in the majority of countries in the Asia-Pacific region (Business Wire, 2018).

5.2.1 Meeting Energy Demand in Asia

According to a report published by Institute of Energy Economics (IEE) Japan, among non-OECD countries, China, India, and the members of the Association of Southeast Asian Nations (ASEAN) will post particularly great growth in energy consumption with their combined growth of 2879 Mtoe,[1] which would amount to Japan's present energy consumption for 6 years.

Fossil fuels, especially natural gas and oil, will be the primary energy sources satisfying the massive energy consumption growth. Although great hopes are placed on nonfossil energy, fossil fuel consumption will increase by 2.3 toe as nonfossil energy consumption

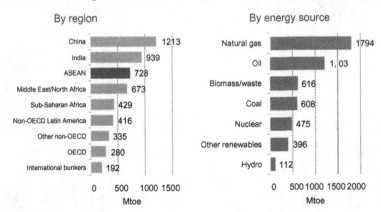

Fig. 5.2 Global primary energy consumption growth (2014–40). *(Reproduced with permission from Asian Development Outlook, 2017. Update—sustaining development through public-private partnership. https://www.adb.org/sites/default/files/publication/365701/ado2017-update.pdf (accessed 8 October 2017).)*

---

[1] Mtoe = million tonnes of oil equivalent.

grows by 1 toe during the outlook period, as shown in Fig. 5.2. Fossil fuels, though reducing their present share of energy consumption from 81%, will still cover 78% of energy consumption in 2040.

5.2.1.1 Association of South East Asian Nations (ASEAN): Primary Energy Consumption

Reflecting strong economic and population growth, ASEAN's primary energy consumption will increase at an annual rate of 3.0% from 624 Mtoe in 2014 to 1352 Mtoe in 2040. The growth will exceed the combined Japanese and Korean present consumption, accounting for 14% of the global energy demand growth. ASEAN will post the third largest growth after China and India.

The ASEAN region, relatively rich with energy resources, is currently a net energy exporter with an energy self-sufficiency rate of 125% (Fig. 5.3). However, as the increase in fossil fuel production will be unable to catch up with the rapid local energy demand growth, the energy self-sufficiency rate will slip below 100% by 2030 and fall to 76% in 2040.

As fossil fuels cover >80% of primary energy consumption growth, ASEAN's rate of dependence on fossil fuels will rise from 74% in 2014 to 77% in 2040. Among energy sources, coal will record the largest growth led by power generation demand, accounting for 34% of

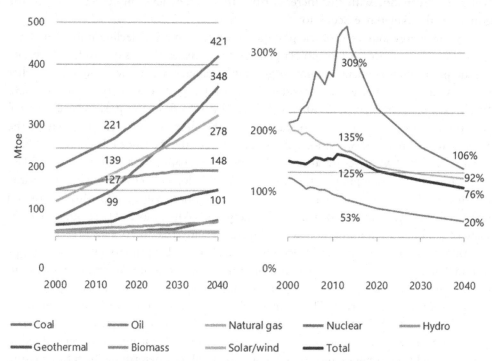

Fig. 5.3 ASEAN primary energy consumption and ASEAN energy self-sufficiency rate. *(Reproduced with permission from Asia/ World Energy Outlook, 2016. Executive Summary. The Institute of Energy Economics, Japan, p. 2.)*

overall primary energy consumption growth. The growth totaling 356 million tonnes of coal equivalent (Mtce) will capture about 40% of the global coal consumption growth.

Oil consumption in 2040 will expand 1.9-fold from 2014, with automobile fuel accounting for half the growth at 4.1 Mb/d. LPG in the buildings sector and petrochemical feedstocks will grow significantly as well. Although oil's share of ASEAN primary energy consumption will fall from 35% to 31%, oil will remain the most important energy source for the region. As ASEAN oil production decreases, the oil self-sufficiency rate will decrease from 53% to 20% as shown in Fig. 5.3.

Excluding the Middle East, East Asia other than China may suffer the largest damage from such source of the supply disruption. Korea and Chinese Taipei would lose GDP as much as their economic growth for some 5 years. Japan would lose as much as growth for about 20 years. Even the European Union with less dependence on Middle Eastern crude oil may suffer an economic contraction of >10%.

5.2.1.2 Advanced Technologies Scenario

The Advanced Technologies Scenario assumes maximum CO_2 emission reduction measures based on their application opportunities and acceptability in society. In that Scenario, energy consumption in 2040 will be 2343 Mtoe or 12% less than in the Reference Scenario, with the increase from the present level limited to 55% of the growth in the Reference Scenario.

Among energy sources, coal will post the largest consumption decline at the Reference Scenario as demand for coal for power generation decreases. This is due to less electricity consumption, improvements on power generation efficiency and fuel switching to other energy sources. Natural gas will register the second largest consumption fall (Fig. 5.3). While coal consumption in 2040 will decrease by 17% from the present level, natural gas consumption will continue to increase for the next quarter century even in the Advanced Technologies Scenario. Oil consumption in the Advanced Technologies Scenario will be 832 Mtoe less than in the Reference Scenario, peaking out around 2040. While fossil fuel consumption in the Advanced Technologies Scenario will be 3196 Mtoe less than in the Reference Scenario, nuclear consumption will be 433 Mtoe more and renewable energy, including solar and wind energy, 419 Mtoe more. As a result, fossil fuels' share of total energy consumption will fall from 81% in 2014 to 70% in 2040.

Although China and India will account for 32% of global primary energy consumption in 2040, their share of energy consumption decline from the Reference Scenario to the Advanced Technologies Scenario will amount to 36%, indicating the two giant Asian energy consumers' great role. They will account for as much as 61% of a coal consumption fall and for 38% of nuclear and wind/solar energy consumption growth. The presence or absence of proactive energy conservation and carbon emission reduction in non-OECD and other countries, rich with potential, will determine the future picture of the world (IEEJ, 2016).

5.2.2 Access to Energy in Asia

Shyamala K. Mani
National Institute of Urban Affairs (NIUA), India Habitat Centre, New Delhi, India

According to Asian Development Bank (ADB, 2009) report, access to modern, cleaner energy is essential to human development. The majority of the world's energy poor are living in Asia and the Pacific. >700 million people in the region have no access to electricity and almost 2 billion people burn wood, dung, and crop waste to cook and to heat their homes.

Energy is crucial to the success of the Millennium Development Goals, and goals that will follow them after 2015. While no Millennium Development Goal focuses on energy, access to it underpins progress toward other goals. Affordable energy decreases poverty and improves quality of life. Modern energy also lifts the burden of women and children whose household responsibilities include spending hours collecting fuel and performing manual labor. Modern energy allows them to pursue more productive activities and can open up opportunities for education. Successful models for increasing energy access need to be replicated increased on a broader scale with the help of the private sector and development organizations (ADB, 2009).

India has power plants with a capacity to generate 300 GW. These are operating at 64% capacity because of the inability of state distribution utilities to purchase electricity and sluggish economic growth. India will not need any new power plants for the next 3 years, as it flushes with the generation capacity, according to a government assessment (IEA, 2009).

The country can manage for the next 3 years with existing plants that are currently underutilized, and those that are under construction, and upcoming renewable energy projects, as shown in the assessment made by the power ministry for reviewing the National Electricity Policy. Table 5.1 ahead illustrates the number and share of people without access to electricity in different states in India (NSSO, 2016; CEA, 2017–18; IEA, 2014). Fig. 5.4 shows the current status of household electrification achieved as of December 2018 under Government of India's *Saubhagya* scheme.

However, ironically, more than a third of the country's population still lives without power. That is because one-third of people in India do not have access to power.

Energy access, according to the Asian Development Bank (ADB) (2009) accounts for the provision of modern energy supplies like electricity, gas, liquid fuels, and other alternatives and the ability of the individuals to purchase these supplies to meet their daily needs. This includes both electricity access as well as access to modern fuels for residential uses, in turn, reducing the reliance on traditional biomass. Being a contributor to sustainable development, sustainable energy should be accessible for all. The Global Tracking Framework (GTF) however, distinguishes the modern biofuels from traditional biofuels, on grounds of sustainability. Traditional biofuels primarily consist of fuelwood, animal dung, and agricultural residues.

Table 5.1 Number and Share of People Without Access to Electricity by State in India, 2013

States	Population Without Access (Million)			Share of Population Without Access		
	Rural	Urban	Total	Rural (%)	Urban (%)	Total (%)
Uttar Pradesh	80	5	85	54	10	44
Bihar	62	2	64	69	19	64
West Bengal	17	2	19	30	7	22
Assam	11	0	12	45	9	40
Rajasthan	10	0	11	22	2	17
Odisha	10	0	11	32	4	27
Jharkhand	8	0	9	35	4	27
Madhya Pradesh	7	1	8	16	3	12
Maharashtra	6	1	6	11	2	7
Gujarat	2	2	3	7	6	6
Chhattisgarh	2	0	3	14	6	12
Karnataka	1	0	1	5	1	3
Other states	3	2	6	2	2	2
Total	221	16	237	26	4	19

Data from National Sample Survey Office, 2014. Household Consumption of Various Goods and Services in India 2011–2012. Government of India, New Delhi; CEA (Central Electricity Authority), 2014. General Review 2014. CEA, New Delhi; IEA (International Energy Agency), 2015. India Energy Outlook: World Energy Outlook Special Report. IEA, Paris, France, 29 pp.

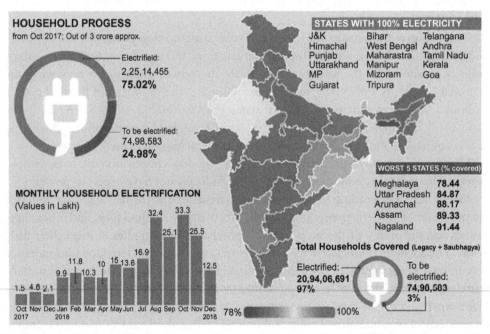

Fig. 5.4 Household electrification by state. *(Courtesy: The Times of India, 2018. Bennett Coleman & Co. Ltd., New Delhi, December 21.)*

Energy from biomass is a significant source of sustainable energy and continues to be one of the largest contributors of renewable energy, especially in Asian countries (WEC, 2010). South Asian countries, in particular, have relatively higher potential, owing to the quantum of agricultural residue as well as MSW available.

Most of the Asian countries still rely on fossil fuels to meet the energy demands and spend considerable amounts of foreign exchange on importing crude oil reserves. Diversifying the energy mix to increase the reliance on renewable energy sources is crucial for enhancing the energy security and mitigation of and formulated policies with a focus on renewable energy-based power sources. Subsequently, biogas is now deemed to provide decentralized electricity solutions in remote rural areas.

The existing and proposed WtE plants could generate renewable energy in an environmentally and economically sustainable manner.

5.2.2.1 Access to Energy in Southeast Asia

In general, countries in Southeast Asia can be divided into three categories. First are the electrified countries—Thailand, Singapore, and Vietnam—in which most households are connected to the grid, and the emerging challenges are energy efficiency and increasing the share of renewables in the energy mix.

Next, come the archipelago countries—Indonesia and the Philippines—where a large majority of the mostly urban population has access to electricity, but connecting the last part of the population is very challenging because they live in remote rural areas or on distant islands.

Finally, there are the energy-poor countries—Myanmar, Cambodia, and Laos—where most of the population is not connected to the grid, rely on kerosene or car batteries for electricity and uses firewood or other biomass for cooking fuel. As the histories, geographies, and resource endowments of these different thematic country groups all differ, splitting them up will help us to understand and analyze the region's access to energy challenges.

Singapore stands alone as the country in Southeast Asia with full electrification and 100% of its population with access to modern cooking fuels. Thailand and Vietnam—with 99.3% and 98% electrification respectively—still have 26% and 56% of their populations relying on traditional biomass for cooking.

However, this will likely change in the next two decades. The share of renewables in the primary energy mix of these countries is expected to fall steeply by 2035, despite large investments and the expansion of solar and hydro capacity. The reason for this steep fall is that the gains in modern renewables will be offset by reduced use of traditional biomass for cooking, as families transition to liquefied petroleum gas (LPG) or other modern fuels. Although this transition has major health benefits, Southeast Asia's energy-related carbon dioxide emissions are also expected to almost double by 2035.

A total of 94% of the Indonesia's urban population is connected, but only 32% in rural areas. However, because they are so populous, are actually home to the majority of Southeast Asia's 127.4 million people who lack access to electricity.

Today, most island communities in Indonesia that do have electricity rely on diesel generators. This reliance is reflected in the country's statistics: 43% of all energy used comes from oil. This pattern is partly the result of history, and partly of convenience. With large oil reserves and an extensive subsidy program, Indonesia has for many years provided cheap and plentiful oil to its citizens. However, these oil reserves are expected to be fully depleted by 2020; already Indonesia is a net importer of oil, and in 2009, it withdrew from OPEC. With domestic oil prices now subject to the unpredictable fluctuations of international trade, even those communities with generators are at risk of poor electricity access. This predicament makes Indonesia a prime candidate for renewable energy-based microgrids although market assessments suggest that current regulatory frameworks are highly unfavorable to off-grid solutions (Discourse Media, 2018).

5.3 WtE TO INCREASE ENERGY ACCESS IN ASIAN DEVELOPING COUNTRIES

5.3.1 Waste to Energy Processes

Shyamala K. Mani
National Institute of Urban Affairs (NIUA), India Habitat Centre, New Delhi, India

Increasing utilization of renewable energy for power generation continues to have a positive impact on the Asia-Pacific WtE market. WtE is the process of generating energy in the form of electricity/heat with the treatment of the waste generated with the use of several technologies such as thermochemical and biochemical. The WtE sector has evolved to generate electricity with the help of various technologies using different categories of waste such as municipal, agricultural, and medical waste, among others.

The energy generated from waste with the help of technologies is used in the form of electricity, fuel, and heat. The WtE management is an important part of the waste disposal infrastructure of the Asia-Pacific region, as WtE is considered an important source of renewable energy. The growth of the Asia-Pacific WtE market is attributed to the rapid industrialization, coupled with the growing demand for renewable energy generation over the forecast period.

However, the high cost of using these technologies in the WtE process, coupled with the complex integrated system, is limiting the growth. To overcome this concern, the governments of different countries are focusing on creating awareness regarding the adoption of these technologies for the conversion process at a larger scale (Business Wire, 2018).

5.3.2 Cost of WtE in Asia

Pratibha Sharma

Global Alliance for Incinerator Alternatives, Pune, India

Most Asian countries offer a striking similarity when it comes to waste produced. >50% is organics or biodegradable, which makes the region's waste the most suited for composting or anaerobic digestion and extremely poor as feedstock for WtE by incineration. Processing organics to be suited for burning requires additional capital and investment, which will increase waste management costs.

In India, the high moisture content in the Indian waste makes it energy-intensive and exorbitantly expensive to burn. In addition, a very efficient informal recycling sector recovers the majority of high-calorific recyclables, thus rendering WtE technologies pointless. A calculation made by the Centre for Science and Environment shows that the proposed WtE plant in Delhi will be able to treat only 10% of the total waste generated in the city. On an average, about 10% of the waste is expected to continue to be recovered for recycling by the informal sector; 65% of it is organics and could be treated using either biological or thermal treatment technologies. The remaining 25% is an inert material that cannot be treated using either biological or thermal treatment technologies (Down to Earth, 61191).

Malaysia has seen several failures of combustion plants in which the calorific value of waste being fed into "waste-to-energy" facilities in certain islands (i.e., Langkawi, Pangkor, Tioman, and Labuan) failed to reach the required heating value. As a result, these "waste-to-energy" facilities were not self-sustaining, and additional fuel had to be injected into each batch of waste treated (Dong et al., 2018).

Thailand through its National Waste Management Plan (2016–21) has been gearing to adopt waste-to-energy approach by encouraging large and expensive investments in this technology. However, these centralized approaches have repeatedly brought governments into disputes with communities and residents living in the targeted areas.

Furthermore, Annexure C of the Stockholm Convention describes waste incinerators as having, "…the potential for comparatively high formation and release of these chemicals [dioxins, furans, PCBs, HCB] to the environment." The ash resulted from waste combustion processing contains varying levels of hazardous toxic chemicals and must be disposed of in landfills that are built to contain hazardous waste.

As an energy plant, it is more expensive than coal or nuclear (EIA, 2013) as shown in Table 5.2. Modern incinerators such as those in the United States and Europe involves the latest air pollution control equipment, regular and frequent emissions monitoring, specialized ash treatment and disposal methods, regular maintenance, and a trained operating crew.

In Bangalore, India, Hanjer Biotech Energies Pvt. Ltd., one of the country's largest solid waste management companies, was revoked permission to set up a solid waste management unit in January 2014 after the company, despite receiving a performance security guarantee of USD 607,693 from the municipality 9 months earlier, neglected their construction responsibilities.

Table 5.2 Capacity and Cost of Waste to Energy Plants in Asia and Other Countries

Locality	Status	Capacity (TPD)	Capital Cost	Capital Cost (US$)	Capital Cost/TPD Capacity (US$)
Dongguan city, China	Unclear	900	US$50,000,000	$50,000,000	$55,600
Shenzen, China	Operating	300	Yuan1.2 billion	$145,000,000	$483,300
Shanghai, China	Approved	1500	US$86 million	$86,000,000	$57,300
Chennai, India	Approved	600	Rs$2000 million	$41,000,000	$68,100
Ringaskiddy, Ireland	Proposed	100	IR pound 75 million	$86,800,000	$868,000
Tokyo, Japan	Operating	400	US$700 million	$700,000,000	$1,750,000
Ibaragi Prefecture, Japan	Operating	160	18 billion yen	$149,100,000	$828,300
Lublin, Poland	Proposed	−375	US$30 million	$30,000,000	$80,000
Ixopo, South Africa	Operating	10	US$60,000	$60,000	$6000
Kwangju, South Korea	Not operating	400	60 billion won	$46,800,000	$117,000
Sanggye-dong, South Korea	Operating	600	80 billion won	$62,500,000	$78,100
Pusan, South Korea	Proposed	200	85 billion won	$66,400,000	$332,000
Suwon, South Korea	Operating	600	90 billion won	$70,300,000	$117,200
Chung Lie City, Taiwan	Approved	1350	NT$4.6 billion	$133,000,000	$98,500
Kaohsiung, Taiwan	Implemented	1800	NT$6.9 billion	$199,500,000	$110,800
Kaohsiung, Taiwan	Implemented	900	NT$3–4 billion	$101,200,000	$112,400
Tainan Town West, Taiwan	Implemented	900	NT$3.8 billion	$109,900,000	$122,100
Phuket Island, Thailand	Operating	250	780 million baht	$17,650,000	$70,600
Tambon Nong Yai, Thailand	Proposed	Unknown	900 million baht	$20,400,000	Unknown
Guam, United States	Proposed	−15	US$13.2 million	$13,200,000	$880,000

TPD, tonnes per day.
Note: Costs have been converted to US$ using August 2001 rates posted on the Universal Currency Converter Website at: http://www.xe.com/ucc.
Reproduced with permission from GAIA's Waste Incineration Database Maintained by Pawel Gluszynski, Waste Prevention Association, Krakow, Poland.

Besides WtE's high investment cost, a particular concern for emerging countries in South and Southeast Asia is its negative impact on waste reduction and diversion efforts, recycling, and greenhouse gas reduction targets. Incinerator companies often utilize a "put-or-pay" contract with cities, forcing cities to either send required a waste volume to a facility or pay the equivalent. Because incinerator contracts typically last at least two decades, the option for cities to implement and embrace waste reduction and resource conservation efforts will be effectively eliminated, as they will be forced to honor these financial contracts to avoid facing legal action from companies.

The Informal Economy Monitoring Study coordinated by the nonprofit WIEGO in five cities in Africa, Asia, and Latin America, found that waste pickers offer a range of economic and environmental benefits as shown in Table 5.3. The table shows the amount of energy conserved as opposed to the energy generated by incineration because of waste collectors' efforts at collecting and sorting dry waste, which is later, recycled. A 2007 GIZ study found waste pickers recovered approximately 20% of all waste material in three of six cities studied (GIZ, 2011). The study found >80,000 people were responsible for recycling about 3 million tons per year of waste across the six cities.

In Delhi, approximately 100,000 waste-pickers recover usable materials such as metal, paper, cardboard, and plastic every day. Altogether, Delhi's recycling economy recovers 1600 tons of waste per day, or approximately 15%–20% of the total generation and saves USD 14,000 per day in operational costs for the city municipality (Tangri and Shah, 2011). As shown in the Fig. 5.5, the informal recyclers also save over 900 thousand metric tonnes of CO_2 equivalent GHG emissions as against mostly less than a 100 thousand metric tonnes of CO_2 equivalent saved by most WtE plants in different cities in India. The Timarpur Okhla plant, which shows saving of 200 thousand metric tonnes of CO_2 equivalent annually is often stalled because of air pollution issues.

5.4 WtE EXPERIENCES FROM ASIA: CASE STUDIES

Roshni Mary Sebastian, Dinesh Kumar, and Babu J. Alappat
Indian Institute of Technology, New Delhi, India

5.4.1 Case Studies in Asian Countries

The waste incineration industry in a few Asian countries, the development, the regulatory framework, emission standards, and the technologies adopted are discussed briefly in the forthcoming sections. Fig. 5.6 shows a few South and East Asian countries and their daily urban MSW generation rates (Hoornweg and Bhada-Tata, 2012a,b; Seo, 2013).

While it is our intention to focus on the best practices in WtE in developing countries in Asia, we are first describing the case study in Japan. The main idea is to bring out not only the contrast in waste quality and incineration capacity to highlight the differences in this technology among developed economies and developing ones in Asia but also to

Table 5.3 Energy Conserved by Recycling Versus Energy Generated From Incineration of Dry Material

Waste Stream Materials	Energy Conserved by Substituting Secondary for Virgin Materials (MJ/Mg)	Energy Generated From MSW Incineration (MJ/Mg)
Paper		
Newspaper	22,398	8444
Corrugated cardboard	22,887	7388
Office (ledger and computer printout)	35,242	8233
Other recyclable paper	21,213	7600
Plastic		
PET	85,888	210,004
HDPE	74,316	21,004
Other containers	62,918	16,782
Film/packaging	75,479	14,566
Other rigid	68,878	16,782
Glass		
Containers	3212	106
Other	582	106
Metal		
Aluminum beverage containers	256,830	739
Other aluminum	281,231	317
Other nonferrous	116,288	317
Tin and bi-metal cans	22,097	739
Other ferrous	17,857	317
Organics		
Food waste	4215	2744
Yard waste	3556	3166
Wood waste	6422	7072
Rubber		
Tires	32,531	14,777
Other rubber	25,672	11,505
Textile		
Cotton	42,101	7283
Synthetic	58,292	7283
Others	10,962	10,713

Reproduced with permission from Resources up in Flames: The Economic Pitfalls of Incineration versus a Zero Waste Approach in the Global South, Brenda Platt, Institute for Local Self-Reliance; GAIA, 2003. Global Alliance for Incinerator Alternatives. Waste Incineration: A Dying Technology. p. 32. Available at: http://www.no-burn.org/wp-content/uploads/Waste-Incineration-A-Dying-Technology.pdf.

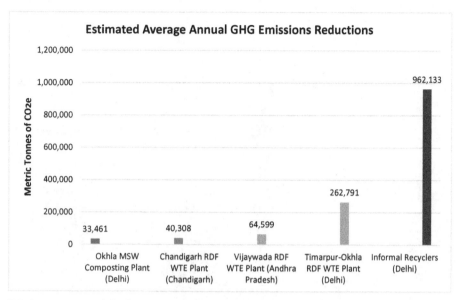

Fig. 5.5 Average annual GHG emissions reductions by waste recovery by informal recyclers in Delhi versus waste incineration plants in different cities in India. *(Reproduced with permission from Cooling Agents, 2009. An Examination of the Role of the Informal Recycling Sector in Mitigating Climate Change. Chintan Environmental Research and Action Group.)*

trace out the history of incineration-based WtE in Asia and the requirement of necessary regulations and safeguards for making it work.

5.4.1.1 WtE Projects in Japan

Incineration as a part of waste management was used in Japan for the first time in 1900 for the control of epidemics. It is now considered as the most sanitary method to treat the wastes in Japan. Japan International Cooperation Agency (JICA) states the primary aims of waste incineration as the reduction of waste volumes to stabilized and neutralized residue (Hershkowitz and Salerni, 1989). With a view to reduce the reliance on fossil fuels and decrease the greenhouse gas (GHG) emissions, a full-scale functioning WtE plant with electricity production was built in 1965 in Osaka by Hitachi Zosen (Hitachi Zosen, 2018). Fig. 5.7 demonstrates the steady growth of the Japanese economy, which consequently led to an increase in the MSW generation. This was met with an increase in the number of waste disposal facilities. As per 2013 records, 1179 incineration plants are functional in Japan (Tabata and Tsai, 2016), with individual plant capacities ranging from <30 TPD to >600 TPD (Ministry of Environment, 2015).

While the number of plants declined by nearly 7.5% from 2009, the individual plant capacities were marginally increased (Yolin, 2015). With incineration of nearly 80% of the MSW generated, Japan has the highest incineration rate as opposed to other countries, including the Organization for Economic Cooperation and Development (OECD)

Fig. 5.6 South and East Asian countries and their daily urban MSW generation rates (in thousand TPD). *(Data from Hoornweg, D., Bhada-Tata, P., 2012a. What a Waste: A Global Review of Solid Waste Management. Urban Development Series Knowledge Papers. World Bank, Washington, DC; Hoornweg, D., Bhada-Tata, P., 2012b. A Global Review of Solid Waste Management. Urban Development Series Knowledge Papers. World Bank, pp. 1–116; Seo, Y., 2013. Current MSW Management and Waste-to-Energy Status in the Republic of Korea, pp. 1–65.)*

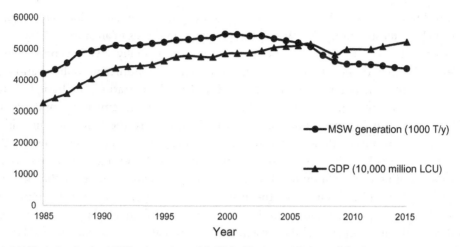

Fig. 5.7 Variation in the MSW generation with GDP. *(Courtesy: Ministry of the Environment, 2015. Solid Waste Management and Recycling Technology of Japan—Towards a Sustainable Society.)*

countries. The collective capacities of the plants amount to 182,683 TPD with an annual power generation capacity of 1770 MW.

Most of the WtE plants in Japan are mass burn units. However, being committed to material recovery, most of the noncombustible fractions are separated from the waste stream prior to incineration. Around 28% of the WtE plants in Japan are dedicated to electricity production (Yolin, 2015) and only a few are earmarked for recovering energy for heating purposes (Tabata and Tsai, 2016). Ministry of Environment (2015) reports that energy from WtE units constitutes only 0.29% of the national energy consumption. The electricity generation and heat production from WtE plants in Osaka were 0.39 and 0.04 kWh/ton, whereas it was 0.41 and 2.57 kWh/ton in an incineration plant in Munich, Germany. This disparity in heat production may be owed to the utilization of heat from WtE plant for district heating. District heating is reported to minimize the CO_2 emissions in comparison to heating of individual buildings. Although many Japanese cities have district-heating networks, only five of them employ heat from WtE plants for the same. One of the reasons cited for this is the low treatment capacity of the plant, which increases the operation and maintenance costs as well reduces the amount of heat produced. Combined with high heat production and supply costs and scanty customers, heat utilization from WtE plants for district heating becomes relatively less favorable in Japan, in comparison to other OECD countries (Tabata and Tsai, 2016).

Regulations for MSW Incinerators in Japan

Japan, the global front-runner in waste incineration promulgated the first legislation pertaining to waste incineration, viz., the Filth Cleansing Law in 1900 to permit waste incineration from a public health perspective. Consequently, cities like Tokyo and Osaka, with limited landfill space established a few incineration units for waste disposal. In 1930, the Waste Cleansing Law was amended to make incineration of MSW mandatory in all municipalities. However, the plants were dogged by poor performance due to inadequate thermal characteristics of the MSW and the pollutant emissions (Yamamoto, 2002).

Toward 1970, a transition to be a more environmentally aware economy was evident in Japan with the establishment and enforcement of numerous policies and legislation. To limit the potential pollutant emissions from incineration, the Basic Law for Environmental Pollution Control, 1967, first established the emission standards for MSW incinerators in Japan. The sources of pollutants were the basis for controlling the potential emissions. Being liable to amendments with scientific knowledge, the standards for SO_2 have been modified 8 times since 1972. With a view to rectifying the short-coming of the "individual sources approach," and account for the ambient air quality, Air Quality Bureau of the Environmental Agency prepared a report in 1974. National Emission Standards in Japan is further supplemented by more stringent standards at the municipal level. For instance, the national emission standard for HCl is 430 ppm; despite this, the emissions from 8 municipal incinerators did not exceed a limit of 100 ppm.

NO_x emissions standards were first formulated in 1973; however, it was not applicable to waste incinerators until 1977. It has been amended 5 times since with technological advancements and largely depends on the age and size of the plant. There exist two national emission standards for particulates in Japan, viz., ordinary emission standards, and special emission standards. The former is applicable all over the country and sets the limits at $0.15 \, g/Nm^3$ [plant capacity > 250 tonnes per day (TPD)] and $0.5 \, g/Nm^3$ (<250 TPD). The latter applies to all facilities that were built post-May 1982 and is located in areas with severe air quality issues. The particulate emission standards are far more stringent according to special standards, with limits defined as $0.08 \, g/Nm^3$ (plant capacity > 250 TPD) and $0.15 \, g/Nm^3$ (<250 TPD). Besides SO_2, NO_x and particulate matter, certain other pollutants were also categorized as toxic emissions, which need regulatory standards as a part of the Air Pollution Control Law. Only HCl standard applies to WtE plants from this and the standards for HCl was established in 1978 as $700 \, mg/Nm^3$.

Following the reports by Olie et al. (1977) regarding the presence of dioxins in the fly ash from the waste incinerator in the Netherlands, Eiceman et al. (1979) detected dioxins in the fly ash from the WtE plant in Kyoto (Olie et al., 1977). Subsequently, the Expert Committee on Dioxins related to Municipal Solid Waste Treatment was created in 1983. The guidelines to Prevent Formation of Dioxins were subsequently issued in 1990 addressing adequate measures to present the formation of dioxins (Yoshida et al., 2009). Adoption of measures to ensure complete combustion and prevent the synthesis of dioxins and removal of dioxins, if any, was proposed to reduce the dioxin emissions to 0.5 ng International Toxic Equivalent (TEQ)/Nm^3. In 1996, the Tolerable Daily Intake (TDI) of dioxins in Japan was set at 10 pg TEQ/kg/d (Ministry of Health and Welfare of Japan, 1996). The facilities with dioxin emissions exceeding 80 ng TEQ/Nm^3 were subjected to emergency measures like closure. The previously formulated guidelines were amended to the Guidelines for Prevention of Dioxin Formation

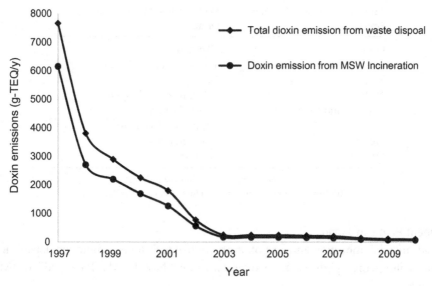

Fig. 5.8 Dioxin emissions from MSW incineration in Japan. *(Courtesy: Government of Japan, 2012. Dioxins.)*

Relating to Waste Treatment in 1997, which mandated the prevention of environmental accumulation.

In consequence of the developments, the Dioxin Risk Assessment Commission adopted 0.1 ng TEQ/Nm^3 as the permanent emission gas criterion. Air Pollution Control Law was modified to include the emission standards for dioxins. It was set at 0.6 pg TEQ/Nm^3 for ambient air. Implementation of various policies also led to an increase in public awareness on the dioxin emission from waste incineration plants. This was met with the sanctioning of Dioxin Special Measure Law, which established the TDI as 4 pg TEQ/kg/d. The objective was to achieve 88% reduction in the dioxin emissions by 2002 from the 1997 emission estimates. After the targeted reduction had been achieved, the plan was later amended to achieve a reduction of 96% from 1998 values by 2010 (Yoshida et al., 2009). Fig. 5.8 shows the reduction in dioxin emissions over the years, because of various technical and policy-level amendments.

The emission standards being followed for MSW incineration in Japan is as shown in Table 5.4.

In 1976, the Waste Disposal and Public Cleansing Law mandated the national government to provide partial financial assistance to municipalities for the construction of solid waste management facilities (Hershkowitz and Salerni, 1989). Prohibition of open burning, integration of construction, and maintenance criteria, etc. was also implemented to improve the financing of WtE plants. Subsidies for installation and maintenance of WtE plants were increased, with special consideration for plants with capacities higher than 100 TPD. Waste Management and Public Cleansing Act was amended in 2010 to further promote waste incineration in Japan (Tabata and Tsai, 2016).

Table 5.4 Emission Standards for MSW Incinerators in Japan

Emissions	Limits
Dust	$40\,mg/m^3$
HCl	$430\,ppm$ or $700\,mg/m^3$
SO_x	$122\,mg/m^3$
NO_x	$250\,ppm$ or $513\,mg/m^3$
Dioxins and furans	$0.1\,ng\ TEQ/m^3$
CO	$30\,ppm$ or $38\,mg/m^3$
Hg	No regulation
HF	No regulation

Note: Reference oxygen concentration is 12% on dry basis.
Courtesy: JEFMA, Japan Environmental Facilities Manufacturers Association.

Prominent Plants in Japan

With advanced combustion and air pollution control techniques employed, Japan's incineration facilities may perhaps, have overcome the "Not In My Backyard" (NIMBY) syndrome.

Shibuya incineration plant was constructed in 2001 within a highly populated urban district in Tokyo. The plant that has been checked for adherence to stringent emission standards for dioxins NO_x, SO_x, and other pollutants has a capacity of 200 TPD. Fluidized-bed technology is used for waste incineration in this facility. Yet another incineration plant was installed in Tokyo Chuo, in 2001 with a capacity of 600 TPD. The plant has a power output of 15 MW. Stoker furnace, fluidized bed and fusion–gasification furnaces are used for incinerating the MSW feed in Japan, with an added benefit of ash recycling. Nearly 70% of the plants in Japan adopt stoker furnaces (Ministry of the Environment, 2012). Fig. 5.9 an illustration of the WtE plants in Japan, located within the city center.

5.4.1.2 WtE Projects in China

China generates 30% of the world's MSW, and the daily generation rates are as high as 533,455 TPD as per 2014 records (Gu et al., 2017). China surpassed the United States of America (USA) to be the largest waste generator and CO_2 emitter in the world in 2004 and 2007 respectively (Hoornweg and Bhada-Tata, 2012a,b). Fig. 5.10 demonstrates the steady increase in waste generation with economic growth in China. Open dumping of wastes was the primary disposal route in China (Hershkowitz and Salerni, 1989), until recently. With limited land availability, adverse environmental impacts of land-filling and increasing GHG emissions from the landfill, waste incineration has emerged as a significant element of integrated waste management in China. Similar to Japan, the primary objective of WtE sector in China is power generation with waste disposal (Zhang et al., 2015).

Fig. 5.9 Shibuya incineration plant, Tokyo and plant in Tokyo Chuo. *(Data from https://tangcatantenna. wordpress.com/2015/06/11/shibuya-incineration-plant/.)*

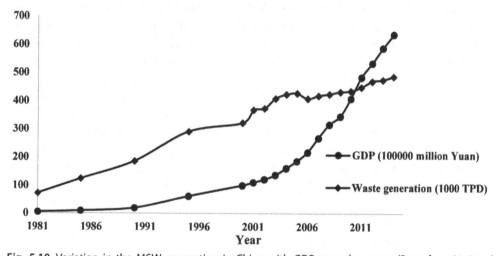

Fig. 5.10 Variation in the MSW generation in China with GDP over the years. *(Data from National Bureau of Statistics of China, 2015. Annual Data. http://www.stats.gov.cn/tjsj/ndsj/2015/indexch.htm (accessed 16 April 2018); Huang, Q.F., Wang, Q., Dong, L., Xi, B., 2006. The current situation of solid waste management in China. J. Mater. Cycles Waste Manag. 8, 63–69.)*

The incinerability of the MSW generated in China has been a concerning factor with respect to the energy efficiency of incineration plants. Being a developing economy, the fraction of organic components was higher, which significantly increased the moisture content to an average of 50% and lowered the heat content to 700–1600 kcal/kg. Combined with poor source segregation and high informal recycling rates, problems like difficulty in ignition, unsteady combustion flame, incomplete combustion of the feed, and increased air pollutant release have been reported. The auxiliary fuel requirement subsequently to ensure complete combustion of the waste feed consequently escalated (Cheng and Hu, 2010; Nie, 2008).

The WtE sector in China is characterized by a relatively late initiation, rapid growth, and high treatment capacities (Xu et al., 2016). Realizing the need to incorporate the novel renewable energy from MSW into the sustainable energy framework in China, the government has formulated favorable policies and stringent emission standards. While the incineration industry had a delayed start in the 1980s, it underwent rapid growth in the 1990s with an approximate treatment capacity of 33,010 TPD by 2006 (Zhang et al., 2010), corresponding to nearly 13% of the MSW generated; this was nearly 6 times that of the installed capacity in 2001 (Chen et al., 2010).

The very first modernized waste incineration plant was built in Shenzen (Guangdong Province) in 1988, with three furnaces of capacity 150 TPD each. Pudong WtE plant of 1000 TPD was established in December 1990 to generate 8.5 MW power. Inclined back and forth ladder mechanical grate was designed to accept waste feeds of calorific value in the range of 1100–1800 kcal/kg. The MSW was stored in a waste pit for 2–3 days prior to feeding to the furnace. To facilitate the complete combustion of the feed, fuel oil was supplied as an auxiliary fuel. Semidry and fabric filter precipitators were employed for air pollution control in the facility and were performed according to European Environmental Protection Standards.

The Dioxins Emission Limit for Chinese MSW Incinerators is 1 ng TEQ/Nm^3 (Liu et al., 2006). By 2016, the number of incineration plants had increased to 220, with a net capacity of 219,080 TPD (China Statistical Yearbook, 2016). The WtE plants are primarily located in the eastern coastal areas, near Yangtze and Pearl River Delta regions. The recently proposed or installed WtE plants are designed with larger capacities of >600 TPD in small cities and as high as 5000 TPD in medium and large cities. Scanty availability of suitable project sites and economic constraints are the prime reasons for designing large WtE plants (Zheng et al., 2014). The largest WtE plant in the world is under construction in Shenzen, China, with a capacity of 5600 TPD, to generate 168 MW power (State of Green, 2018).

Grate furnace and fluidized-bed furnaces are predominantly used in WtE plants in China. Nearly 80% of the WtE plants in China rely on these technologies for incineration of MSW (Ministry of Construction, 2007). Use of reciprocating furnaces in China is mostly confined to special industrial wastes. Grate technology is mostly adopted for plants with an average capacity of 500 TPD and a maximum of 1000 TPD. Fluidized-bed technology is, however, adopted for plants with capacities ranging from 100 to 500 TPD. Moreover, the fly ash generation is reportedly higher in fluidized-bed incinerators; the susceptibility to the erosion of heat transfer furnaces further makes it a relatively unfavorable choice (Yokoyama et al., 2001).

Subsequently, WtE plants in larger cities like Shanghai and Beijing implement stoker-fired furnaces (Liu et al., 2006). Fluidized-bed technology, however, had a quick growth in Chinese WtE sector, owing to relatively lower installation, operation, and maintenance costs. Furthermore, it allows the use of coal as auxiliary fuel to incinerate waste feeds with high moisture content and low calorific value generated in China and subsequently suppressing the dioxin emissions (Gullett et al., 2000; Yan et al., 2006).

Regulations for MSW Incinerators in China

The Chinese government has been dedicated to promoting MSW incineration for waste disposal and power generation. Numerous policies have been formulated to ensure the growth of WtE sector. Renewable Energy Law of the People's Republic of China, enforced in 2006 is the first law to provide legal protection and support for the development of renewable energy sources.

Aiming to specifically cater to WtE sector, it was amended in 2009; this recognized waste incineration as an integral part of national energy resource. The law ensures budget allocations from the central government and collection of renewable energy fees, which eventually paves way for the unhindered development of WtE plants (Zheng et al., 2014). The National Program on Climate Change in China (2007) supports the installation of WtE plants in areas with limited land availability and that are economically developed.

The policy "Opinions on Further Strengthening the Work of MSW Disposal" was approved in 2011 with a view to increase resource utilization by various techniques, including incineration (State Grid, 2018). In 2013, "Opinions on Promoting the Service in distributed Power Parallel to the Grid" was established to offer benefits to promote WtE sector (State Grid, 2018). The tipping fees from the local governing bodies ranged from 13.04 to 19.56 USD/ton and were exempted from tax and value-added tax (VAT) is returned after paying. The plants receive a 100% tax exemption from income tax in the first 3 years and 50% exemption from 4th to 6th year. The emission standards have simultaneously been made stringent to facilitate the healthy development of WtE sector (Xin-Gang et al., 2016). Table 5.5 lists some prominent policies formulated in China to promote the growth of MSW incineration.

Pollution Abatement Measures

Incineration of MSW may tend to be environmentally exhaustive in the absence of adequate control and precautionary measures. The State Environmental Protection Administration of China formulated the Pollution Control Standards in 1999. Like Japan, the local Environmental Protection Bureau (EPB) limits are more stringent in comparison to national limits (see Table 5.6). While the national emission standard for dioxin is $1\,ng$ TEQ/Nm^3, the local EPB demands a much lower limit of $0.1\,ng\,TEQ/Nm^3$. However, in comparison to standards in the USA and Europe, it was considerably lenient in China.

Table 5.5 Regulatory Frameworks Pertaining to WtE Technologies in China

Date	Policy	Objectives
1999	State Environmental Protection Administration of China	Pollution control standards to facilitate the growth of WtE sector
2006	The 11th Five-Year Plan of National Urban Environmental Sanitation	Recommendations for waste disposal in different regions
2006	Renewable Energy Law of the People's Republic of China	Legal protection and phased support for the development of the WtE sector
2007	Municipal Solid Waste Management Approaches	The registered capital of WtE plants is not <16.3 million USD
2007	The State 11th Five-year Plan on Facility Construction of Urban Waste Harmless Disposal	Encouraging WtE plants; increase the rate of incineration in Eastern cities to at least 35%
2007	National Programme on Climate Change in China	Installation of WtE plants in areas with limited land availability and is economically developed
2008	The Notice on Improving the Assessment and Management of Environmental Impact of Biomass Power Projects	Environmental protection distance of new project expansion shall be at least 300 m
2008	Notice on Comprehensive Utilization of Resources and the VAT Policy of Other Products	VAT returned after paying for WtE projects
2009	Directory of Environmental Energy and Water Conservation Projects Preferential Corporate Income Tax	Preferential corporate income tax for WtE projects
2011	The State 12th Five-Year Plan of Environmental Protection	Encouraging various waste treatment techniques, including waste incineration
2012	Notice About Improving Waste Incineration Power Price Policy	WtE benchmark price of 0.106 USD/kWh
2013	The 12th Five-Year Plan of National Environmental Standards	Revision of emission standards for WtE plants
2013	Opinions on Promoting the Service in Distributed Power Parallel to the Grid	Selective benefits to WtE sector
2014	Standards for Pollution Control of Municipal Solid Waste Incineration	Stringent emission standards comparable with international standards formulated

Data from Ministry of Environmental Protection, 2014. Executive Summary of Environmental Assessment for China GEF Municipal Solid Waste Management Project. Foreign Economic Cooperation Office.

Generation of hazardous fly ash is yet another alarming factor, since 2 million T/yr of fly ash is generated by MSW generation. Stoker incinerators generate 3%–5% and fluidized-bed incinerators generate 10%–20% of fly ash from the MSW feed. Although fly ash needs to be disposed of in hazardous waste landfills, it is rarely practiced and gets unscientifically dumped. Studies to use sintering technology to convert fly ash into nontoxic products and use it to produce construction materials are being carried out. A 99.3% decomposition of dioxins was obtained in a pilot study of sintering process (Nie, 2008).

Table 5.6 Emission Standards for MSW Incinerators in China

Emission	Limits
Particulates	$80 \, mg/m^3$
Gas darkness	1 Lingleman, grade
SO_2	$260 \, mg/m^3$
NO_x	$400 \, mg/m^3$
CO	$150 \, mg/m^3$
HCl	$75 \, mg/m^3$
Hg	$0.2 \, mg/m^3$
Cd	$0.1 \, mg/m^3$
Pb	$1.6 \, mg/m^3$
Dioxins	1 TEQ ng/m^3

Note: Reference oxygen concentration is 11% on dry basis.
Data from Ministry of Environmental Protection, 2014. Executive Summary of Environmental Assessment for China GEF Municipal Solid Waste Management Project. Foreign Economic Cooperation Office.

Prominent Plants in China

The largest circulating fluidized-bed incineration plant in China with a capacity of 800 TPD was installed in 2012 in Cixi, based on the technology developed by Zhejiang University. Provided with air preheaters and shredders to ensure combustion efficiency, this plant can incinerate MSW with calorific value even <1000 kcal/kg without auxiliary fuel (Huang et al., 2013). Table 5.7 lists the prominent WtE plants in China.

5.4.1.3 WtE Project in India

India is yet another rapidly developing country as China with an astonishing growth rate of population and waste generation. Nearly 62 million tonnes of urban MSW are generated in India annually, most of which ends up unscientifically disposed of in landfills (Planning Commission of India, 2014; Hoornweg and Bhada-Tata, 2012a,b). The steady increase in population and economic growth has led to an astounding increase in the MSW generation, as demonstrated by Fig. 5.11. The gradual yet steady growth in Chinese and Indian WtE sector has made the Asia-Pacific region the fastest growing market for WtE techniques (World Energy Council, 2016). The estimated energy potential for MSW in India is 1.5 GW of which only 2% is currently utilized (MNRE, 2013).

Many of the incineration plants in India have not been successful owing to erroneous MSW quantity and quality data, poor coordination between the plant operators and the ULBs and financial unsustainability. The failure of 300 TPD incineration plant in Timarpur had decelerated the growth of WtE sector initially. The WtE plants in Kanpur, Elikatta (Andhra Pradesh), Vijayawada, and Rajahmundry are presently nonoperational. The improvement in the thermal characteristics of the MSW has triggered an increase in the number of plants being commissioned.

Table 5.7 Prominent Incineration Plants in China

Year of Construction	Name	Capacity (TPD)	Power (Million kWh)	Investment (USD)	Technology
1988	Shenzhen Qingshui river MSW incineration plant	300	–	–	–
2000	Qiaoshi incineration plant, Hangzhou	800	–	–	Circulating fluidized bed
2001	Ningbo, Zhejiang province	1000	–	–	Ladder mechanical grate
2002	Shanghai Pudong MSW incineration plant	1000	100	110	Stoker grate
2005	Shanghai Jiangqiao waste incineration power plant	1500	180	144	
2011	Shanghai Jiangqiao waste incineration power plant	2000	270	147	
2013	Guangzhou Likeng second MSW incineration plant	2250	290	152	
2013	Beijing Lujiashan MSW incineration plant	3000	310	329	

Data from Ministry of Environmental Protection, 2014. Executive Summary of Environmental Assessment for China GEF Municipal Solid Waste Management Project. Foreign Economic Cooperation Office.

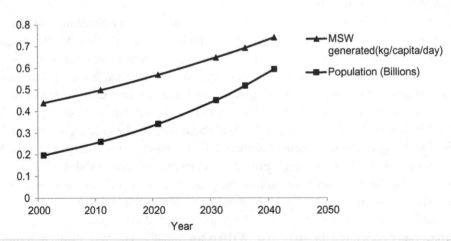

Fig. 5.11 Per capita MSW generation rates and the population growth in India. *(Reproduced with permission from Sebastian, R.M., Alappat, B., 2016. Thermal properties of Indian municipal solid waste over the past, present and future years and its effect on thermal waste to energy facilities. Civ. Eng. Urban Plan. An Int. J. 3, 97–106. doi:10.5121/civej.2016.3208.)*

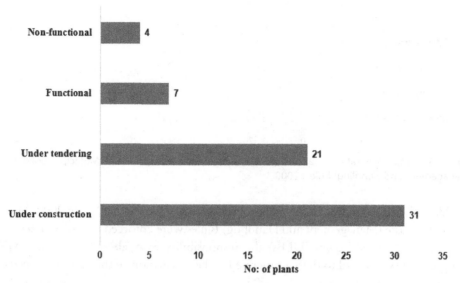

Fig. 5.12 WtE plants in India.

Fig. 5.12 gives an approximate number of WtE facilities in India, in different stages of construction/installation. Table 5.8 lists a few prominent MSW and RDF incineration plants operational in India now.

Regulatory Framework

Ensuing from a multitude of public interest litigations the Supreme Court of India, in 1996, directed the Ministry of Environment and Forests (MoEF) to formulate the rules

Table 5.8 WtE Plants Operational in India

Location	Capacity (TPD)	Power Generation (MW)	Technology
Okhla	1950	16	Mass burning; now changed to RDF
Ghazipur	1300	22	RDF
Village Mandur, Bengaluru	1000 (not operational)	8	–
Nalgonda, Andhra Pradesh	–	11	RDF
Hadapsar, Pune	300	10	Gasification
Jabalpur, Madhya Pradesh	600	8.5	Mass burning
Narela Bawana, Delhi	3000	24	RDF
Karimnagar, Andhra Pradesh		12	Mixed feed

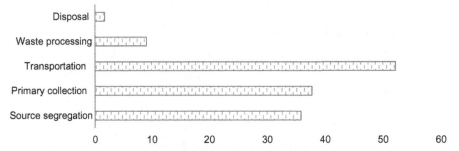

Fig. 5.13 Compliance rate of various MSW management activities with Municipal Solid Waste (Management and Handling) Rules, 2000.

to address the concerns due to the inappropriate management of MSW. In 2000, Municipal Solid Waste (Management and Handling) Rules were enforced in Indian states assigning the urban local bodies (ULBs) the responsibility to establish an efficient MSW management system and to draft a timeline for the installation of the various processing, treatment, and disposal facilities by 2003.

While the rules stated guidelines for the collection and transportation of wastes, no restrictions on the type of treatment technique were imposed. However, the targets could not be achieved by 2003 and the compliance to the rules was found to be unsatisfactory (Fig. 5.13). This was primarily due to the poor financial support as well as lack of technical expertise to implement the technologies at ULB level (Dube et al., 2014). The Ministry of Urban Development (MoUD) to ensure financial assistance was given to the ULBS for infrastructural projects launched the Jawaharlal Nehru National Urban Renewal Mission (JNNURM) in 2005, with special emphasis on MSW management projects (Table 5.9; MoEFCC, 2016).

Table 5.9 Emission Standards for MSW Incinerators in India

Parameters	Concentration (11% O_2 Correction on Dry Basis)
Particulate matter	$50\,mg/Nm^3$
Nitrogen oxides	$400\,mg/Nm^3$
HCl	$50\,mg/Nm^3$
SO_2	$200\,mg/Nm^3$
CO	$100\,mg/Nm^3$
Total dioxins and furans	$0.1\,ng\ TEQ/m^3$
Cd + Th + their compounds	$0.05\,mg/Nm^3$
Hg and its compounds	$0.05\,mg/Nm^3$
Sb + As + Pb + Cr + Co + Cu + Mn + Ni + V + their compounds	$0.5\,mg/Nm^3$

Data from Ministry of Environment, Forest and Climate Change, 2016. Solid Waste Management Rules. Government of India. http://www.moef.nic.in/content/so-1357e-08-04-2016-solid-waste-management-rules-2016?theme=moef_blue (accessed 17 April 2018).

The scheme also invited the involvement of the private sector for financial and technical innovations. The 12th Finance Commission sanctioned nearly 415 million euros for reinforcing MSW management activities. The 13th Finance Commission recommended that 50% of the funds received by the ULBs be allotted for MSW management. According to the National Implementation Plan for Persistent Organic Pollutants (PoPs), 94% of the MSW was reportedly dumped, 4% composted, and 2% recycled (Government of India, 2011).

Yet another distressing factor is the emission standards for incinerators in India. The emission standards specified by the MSW (Management and Handling) Rules, 2000 are far lower in comparison to the US or European limits. The rules also failed to establish any standards for toxic emissions like dioxins and furans, heavy metals and mercury.

During 2003–05, MoEF in collaboration with National Institute for Interdisciplinary Science and Technology (NIIST) monitored emissions from medical waste incinerators as an initial research on dioxin emissions. National Environmental Engineering Research Institute (NEERI) and many other national laboratories had also taken up research into the dioxin emissions from waste incinerators. The National Mission on Sustainable Habitat (NMSH), 2010, under the aegis of National Action Plan on Climate Change (NAPCC) came into effect to promote energy efficiency and effective solid waste management through amendments in the legal and policy framework. The Central Public Health and Environmental Engineering Organization (CPHEEO) under MoUD later released a manual with a view to aiding the decision-makers in effectively managing the MSW in urban areas. The Ministry of New and Renewable Energy (MNRE) is also actively involved in promoting renewable energy production. The National Master Plan (NMP) for Development of WtE in India aimed to create a conducive environment for utilization of urban MSW for energy recovery and the installation of WtE plants for its effective utilization.

Financial assistance of up to 3 crores or 50% of the project costs is granted to eligible WtE projects under the NMP (Dube et al., 2014). The Swatch Bharat Mission launched by the MoUD in 2014 also offers financial assistance to WtE projects. Options to receive central government grants are either directly or as a generation-based incentive.

Furthermore, to address the shortcomings of the existing rules, the Government of India formulated the Solid Waste Management Rules, 2016. Swacch Bharat Mission in collaboration with various Ministries and Departments intend to promote Waste to wealth and enforce the Solid Waste Management Rules, 2016 (MoHUA, 2017). The rules clearly emphasize the source segregation of the wastes and mandates only the inert and noncombustible resides to be landfilled. The emission standards for MSW incinerators were also amended to account for dioxin and furan emissions (MoEFCC, 2016).

Fig. 5.14 illustrates two presently operational WtE plants in India.

The WtE plants currently under construction or at the tendering stage, as per Ministry of New and Renewable Energy (MNRE) database, are given in Tables 5.10 and 5.11.

Fig. 5.14 (A) Timarpur-Okhla WtE plant, Delhi. (B) Waste storage pit in Jabalpur Plant, India.

5.4.1.4 WtE Projects in Indonesia

Shyamala K. Mani
National Institute of Urban Affairs (NIUA), India Habitat Centre, New Delhi, India

As a holistic project on solid waste management and for examining the possibility of converting waste to energy, Urban Nexus project coordinated by GIZ, took up a series of case studies in Vietnam, India, Indonesia, and the Philippines from 2017. In Indonesia, the Tanjung Pinang Municipality was selected under this project for improving the solid waste management system, its landfill, and examined it for the possibility of producing energy from waste.

Tanjung Pinang is the capital and the second largest city of Riau Islands province. It has an area of $239.5\,km^2$ and a population of 229,396 inhabitants. Tanjung Pinang is located south of Bintan Island and has ferry connections to Batam (Indonesia), Singapore, and Johor Bahru (Malaysia). The average daily production of waste is 400 TPD, of which 160 TPD is collected. The landfill, Ganet Sanitary Landfill is 10.8 ha in area. An expert from the Chaiangmai Sanitary Landfill, examined the possibility of extracting methane from the landfill, which can be used as a fuel for energy generation.

Due to the shortage of electricity in the Tanjung Pinang Island electricity was going to be imported from Batam Island, which would have been very expensive. Instead, electricity from Landfill Gas from the waste in the Ganet Landfill was an option that was explored while the possibility of using it as compressed biogas (CBG) was also examined especially because it required less investment than LFG to electricity. Furthermore, its product replaced precious fuel like diesel. Since this process of extraction of gas from the landfill, requires a lot of land, MYT or the Maximum Yield Technology, which requires less land and is efficient for recycling and recovering 90% of the waste was also looked at. The study on Solid Waste Management using MYT technology was completed in April 2017.

The Landfill has been operational since 1999 until the present, zone 1 to zone 3 are closed cells and zone 4 is an operating cell. A small-scale LFG collection system is already

Table 5.10 List of WtE Plants Under Construction in India

Sl. No.	State	City and Capacity (MWe)	
1	Andhra Pradesh	Vishakhapatnam	15
		Tirupati	5
		Nellore	4
		Kurnool	1
		Vizianagaram	4
		Vijayawada	12
		Kadapa	5
		Anantpur	4
		Kurnool	1
		Tadepalligudem	5
2	Bihar	Patna	12
3	Gujarat	Surat	11.5
		Rajkot	4
4	Himachal Pradesh	Shimla	1.7
5	Jharkhand	Ranchi	11
6	Karnataka	Bengaluru	8
		Bengaluru	12
7	Kerala	Kochi	10
8	Madhya Pradesh	Indore	8
9	Maharashtra	Pune	7
		Thane	10
		Nagpur	11.5
10	Manipur	Imphal	1
11	New Delhi	Kidwai Nagar	1
12	Odisha	Bhubaneswar and Cuttack	11
13	Punjab	Ludhiana	8
		Bathinda	8
14	Tamil Nadu	Coimbatore	8
15	Telangana	Greater Hyderabad Municipal Corporation	11
16	Uttar Pradesh	Allahabad	6
		Agra	10
			Total 241.8

Data from Ministry of New and Renewable Energy (MNRE), 2017. http://www.indiaenvironmentportal.org.in/files/file/Waste-%20to%20-Energy%20Plants.pdf (accessed 20 July 2019).

operating and is able to produce electricity via a small generator just enough to power a few light bulbs in the landfill buildings. The system also uses the collected LFG for occasional cooking. The small-scale waste-to-energy project shows that Tanjung Pinang is on the right track in their environment and resource management. However, it was thought that the TPA landfill gas has much more potential in terms of energy production from LFG.

Based on the study, in order for the TPA landfill to become a sustainable sanitary landfill and reach its highest potential in landfill gas collection and utilization, it was

Table 5.11 WtE Plants Under Tendering Stage in India

Sl. No.	State	City	Capacity (MWe)
1	Andhra Pradesh	Guntur	15
2	Assam	Guwahati	5
3	Chhattisgarh	Durg-Bhilai	5
		Raipur	5
4	Gujarat	Ahmedabad	15
5	Haryana	Karnal	3.5
		Sonepat	5
		Bandhmadi	10
6	J&K	Sir Nagar	6.5
7	Jharkhand	Dhanbad	12
8	Madhya Pradesh	Bhopal	9.5
		Rewa	5
		Ujjain	3.5
		Gwalior	6
9	Rajasthan	Jaipur	15
		Kota	5
		Jodhpur	6
10	Uttar Pradesh	Rampur	3
		Gorakhpur	3
		Jhansi	3
11	West Bengal	Kolkata	22.5
			Total 163.5

Modified from Ministry of New and Renewable Energy (MNRE), 2017. http://www.indiaenvironmentportal.org.in/files/file/Waste-%20to%20-Energy%20Plants.pdf (accessed 20 July 2019).

recommended that the TPA landfill adopt Bantan's "Sustainable Sanitary Landfill to Energy System (SSLTES)." This SSLTES concept allows for the excavation of the closed cell (after 15–20 years when all the organic waste has completely decayed). The excavated cell becomes a "new" cell for dumping. By applying this method the excavation can be done in a series (one cell after another), and therefore, it was envisaged that Tanjung Pinang landfill would never run out of land to dump garbage. The SSLTES concept also decreases open space during operation. The only open space at any given time would be the area opened for garbage dumping.

The rest of the space would either be covered by HDPE, temporary plastic sheets, or daily cover (soil). By applying this method, less odor will be released into the atmosphere, less spaces for insects to use for colonization, as well as less leachate due to less rain penetration into the landfill. As a result, the nearby communities would be less affected by the landfill.

As electricity can be produced from LFG, Tanjung Pinang would benefit from larger scale biogas collection and electricity production. At least 1 MW could be produced from 200 to 400 tons of solid waste per day. The electricity produced will contribute to the needed additional source of energy.

Tanjung Pinang in cooperation with GIZ Nexus, therefore, embarked on Sanitary Landfill: Solid WtE Project. The Preliminary Feasibility Study of implementing Sustainable Sanitary Landfill to Energy System (SSLTES) in Tanjung Pinang was elaborated and presented to the local Authorities. The study focuses on the following:

- Possible application of new landfill management concept: landfill preparation, pipe and drainage installation, daily cover methods, and dumping and compacting methods.
- Capturing of LFG (Land Fill Gas) and capturing gas from leachate treatment with appropriate and efficient technology.
- Production of either electricity from biogas.
- Investment cost estimation.

The two possibilities of energy type to be produced are Electricity or Compressed Bio Gas (CBG) as fuel for vehicles. Investment to produce Electricity would cost 27.3 Billion IDR (1.8 Million EUR) and the return of investment period is 6 years and 5 months. Moreover, Tanjung Pinang is facing power shortage and blackout occurs often daily. Therefore, any additional power would be welcome. Investing in CBG production could be financially viable as it is relatively less investment than electricity production. Moreover, the total amount of LFG collected can produce 6 ton/day of CBG, which is equivalent to 4620 L/day of diesel.

In an alternative approach cited in an article by Martin Schaub and Jérémie Bertrand, a concept for the implementation of a waste management system in the city of Tanjung Pinang, Indonesia that can be realized step-by-step was also described. This sequential approach allows the city of Tanjung Pinang to develop and improve their waste management system over time and in line with its financial possibilities. In order to improve the waste management system in Tajung Pinang, the report developed a step-by-step strategy that in its final stage leads to the implementation of mechanical biological treatment (MBT) at Ganet Landfill. In an MBT plant, waste is turned into compost or biogas and refuse-derived fuel (RDF) which can be used for power generation. Hence, an MBT plant allows making maximum use of the energy in the waste.

The project team is convinced that the implementation of a mechanical biological waste treatment process on the Ganet Landfill site constitutes the best solution also in terms of rationalization of costs (workforce, gas treatment, and water treatment). This report emphasizes the necessity to implement a waste management system in phases consisting of mechanical pretreatment, anaerobic digestion/MYT, and subsequent composting/bio drying.

However, the progress of the project after a year or so, published by WEHRLE in February 2018, on the Nexus project at Tanjung Pinang, Indonesia, as far as the landfill gas capture goes, says that there is already some preliminary infrastructure to capture the methane and use it for cooking or flaring. The report says that the consequent expansion and capture of landfill gas should be the plan at Ganet landfill. The report also says that similarly for leachate treatment at the landfill, there is a small leachate treatment plant, which is not state of the art and treats the leachate only minimally. It has recommended upgrading the leachate treatment plant to enable the treated leachate to adhere to the national standards. The report also cites the existence of a simple composting plant, which, again it says needs to be upgraded and expanded. According to the report the MBT and MYT (Maximum Yield Treatment) have yet to be achieved as per the description in the Nexus Case study plan (Aritono, 2017).

5.4.2 Biomethanation (Anaerobic Digestion) Systems for MSW to Energy in Asian Countries

Suneel Pandey and Dinesh Chandra Pant
The Energy and Resources Institute (TERI), New Delhi, India

Increase in MSW generation is mainly due to rapid urbanization, rising consumption patterns, the related increase in MSW generation, change in waste characteristics over the years and lack of awareness and public apathy toward the seriousness to deal with the issue. MSW is a highly variable and heterogeneous, multicomponent material, which varies both seasonally and geographically. Bulk of this waste is being dumped in the open in an uncontrolled manner resulting into the pollution of water bodies and land and causing uncontrolled emission of methane.

The calorific value of the MSW is contributed to biodegradable content of the waste such as food waste and nonbiodegradable content such as paper waste, plastics, rags, leather, etc. The characteristics and variability of MSW as a fuel has a significant impact on its behavior as a fuel in combustion and other thermal processing systems. In addition to the variability in composition, MSW is notoriously difficult to handle and to feed in a controlled manner, in incineration equipment. This is reflected in the design of MSW handling and feeding systems and has a significant knock-on effect on the difficulties encountered in the control of the combustion conditions in a conventional incineration plant. MSW is also a high slagging and fouling fuel, that is, it has a high propensity to form fused ash deposits on the internal surfaces of the furnace and high-temperature reactor, and to form bonded fouling deposits on heat exchanger surfaces.

5.4.2.1 Biomethanation Systems for MSW in Asia

Solid and liquid wastes consist of both organic and inorganic constituents, and the degradation of the former can take place in the presence or absence of oxygen (air). When microbial degradation of organics takes place in the absence of air, the process is known as "anaerobic digestion" or "biomethanation." This results in the production of biogas, which contains methane, carbon dioxide, and traces of other gases. Anaerobic digestion occurs naturally in swamps, waterlogged soils and rice fields, deep-water bodies, and in the digestive systems of animals. Anaerobic processes can take place in a reactor such as a digester vessel, covered lagoon or landfill in order to recover the methane gas (as biogas), which can be used for power generation, thermal application and/or automotive fuel as Bio-CNG (bio-compressed natural gas). Further details are found in Chapter 3 of this book.

Biomethanation systems are among the most mature and proven processes, which converts waste into energy efficiently and can be used to achieve the following goals:

- Pollution prevention/reduction.
- Reduction of uncontrolled Greenhouse gas (GHG) emissions and odor.
- Recovery of bioenergy potential as biogas for fuel/electricity generation.
- Production of stabilized residue for use as manure.

Commercially, biomethanation projects have proven their viability under regimes of user charges or tipping fee in various countries in Europe and the United States by contractors putting up such projects. Unlike Advanced Thermal Conversion Technologies (ATC), relatively few patented biomethanation technologies are in existence. However, biomethanation is a mature technology and there is extensive experience worldwide in all the aspects of biomethanation. Thousands of biomethanation plants have been built, ranging from small farm digesters to large-scale waste treatment and biogas recovery plants. Biomethanation technologies are among the most promising options for WtE recovery, particularly for agriculture-based countries such as India. Source separated MSW has been used as a feedstock in the existing plants. The country with the greatest experience using large-scale digestion facilities has been Denmark, where 18 large centralized plants are now in operation. In many cases, these facilities co-digest manure, clean organic industrial wastes, and source separated MSW. Commercially available technology providers— Kruger (Denmark), Kompogas (Switzerland), Entec (Austria), Eco Tech (Finland), BTA (Germany), and Dranco (Belgium) are some of the major players who have installed most of the units around the globe.

As mentioned above, there are not too many biomethanation technologies that have been patented. However, various distinctive processes and unique technologies have been developed and are being used for biomethanation of organic waste. Some technologies are listed below in Table 5.12.

In India, early biogas digester use is recorded for 1859 when it was used in sewage treatment. The technology since then has been widely used in Asia mainly for processing cattle dung with often small amount of food or kitchen waste. The historical development of biogas technology in India until 1950–60 is shown in the Fig. 5.15.

The development drivers for introducing such systems to provide people with biogas is to reduce consumption of firewood and the respective deforestation, decrease indoor air pollution and improve soil fertility. After roughly 25 years of stepwise improvements and practical experience, the technology is still attracting interest as a contribution to renewable energy production and creating independence from fossil fuels. The Ministry of Agriculture, China, added an estimated 22 million biogas systems between 2006 and 2010 to reach a total of some 40 million installed systems in early 2011. India is home to

Table 5.12 Biomethanation Technologies

Types of Digesters	Type of Fermentation/ Digestion	Technologies
Single-stage	Low-solids or wet	The Wabio process
		The Waasa process
		The BIMA digester
	High-solids or dry	The Dranco process
		The Kompogas process
		The Valorga process
Multistage	Continuous dry-wet	The BTA process
		The Linde process
		The Arrowbio process
		The Biopercolate process
	Continuous wet-wet	TERI's enhanced acidification and methanation (TEAM) process
Batch systems	One stage	The SEBAC system
		The anaerobic phased solids digester

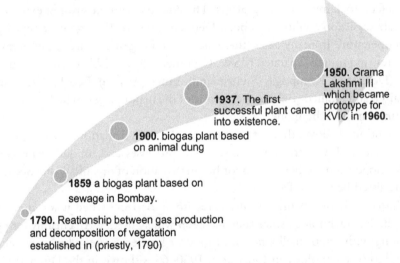

1950. Grama Lakshmi III which became prototype for KVIC in **1960.**

1937. The first successful plant came into existence.

1900. biogas plant based on animal dung

1859 a biogas plant based on sewage in Bombay.

1790. Reationship between gas production and decomposition of vegatation established in (priestly, 1790)

Fig. 5.15 Historical development of biogas in India.

approximately 4.94 million systems by 2017, and Vietnam has installed 20,000 systems annually to reach >100,000 by 2010. Cambodia, Laos, and Indonesia have smaller biogas programs, nevertheless installing about 1000 systems in each country in 2010. Nepal's Biogas Support.

Program, which involves the private sector, microfinance organizations, community groups, and NGOs, has resulted in a steady increase in installed biogas systems during the last decade. Approximately 25,000 systems were constructed in 2010, bringing the nationwide total to nearly 225,000 (REN21, 2011).

Recycling of organic waste would significantly reduce the amount of waste that needs handling and thus reduce costs at the disposal facilities. Less organic waste at the disposal site prolongs its life span and reduces the environmental impact of the disposal site as the organics are largely responsible for the polluting leachate, methane, and odor problems. The implementation of anaerobic digestion or composting as one step in a city-wide solid waste management programme reduces the flow of biodegradable materials to landfills. Nevertheless, its feasibility depends on the market demand for the end products (gas, digestate or compost), as well as the technical and organizational setup of the individual facilities. Enabling clear legislation, policy, and municipal strategy in terms of organic waste management are further important prerequisites for successful initiatives (Zurbrugg et al., 2004).

In most low- and middle-income countries separate collection of household segregated organic waste is rare. Thus, the collected waste consists of organic waste mixed with other waste materials and any planned organic waste treatment will require subsequent sorting of the biowaste fraction. This not only leads to additional costs but more importantly results in lower quality biowaste feedstock. An exception to this is waste collected from sources, which generate predominately biowaste with few contaminating inorganic substances, such as wastes from vegetable markets, restaurants or food processing industries. Biogas plants should, therefore, be built or located where large amounts of organic waste are accumulated.

Source segregated biowaste is generally of higher quality as it contains less nondegradable contaminants such as glass, plastic, rubber, stones, sand, and hazardous and/or toxic substances. Such comparatively "pure" biowaste is thus ideal for treatment in an AD system. Manual sorting is nevertheless inevitable to ensure that impurities in the feedstock to an AD facility are removed, as this may lead to clogging of inlet pipes, reduced biogas yield, and lower quality and acceptability of the digestate.

It is important to note that the storage time of collected biowaste should be as short as possible, especially in hot and humid climates. During storage, organic matter starts to decompose. Thus, with longer storage biogas yield will decrease as waste has already degraded and lost some of its energy value. High amounts of fibrous material (e.g., straw) or clumps of material need to be avoided as this hinders the degradation process in the digester. Straw can cause considerable scum layer formation, which is difficult to control during digestion. Breaking apart the clumps in a rotating drum and shredding is recommended. Reducing the particle size of the feedstock is important to avoid blockage of the inlet pipe, and to increase ease of degradation. As a general rule, substrate particle size with a diameter of max. 5 cm is recommended although the ideal size also depends on the diameter of the inlet pipe. Shredding of the feedstock into small particles increases the total surface area of the material thus increasing the area that can be degraded by microorganisms (Schnürer and Jarvis, 2010), as many microorganisms, especially those that are

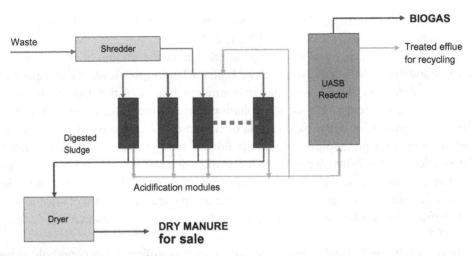

Fig. 5.16 Schematic diagram of TEAM process.

active in the initial hydrolytic step, prefer to attach to the surface area of the material that they are degrading.

The choice of the basic AD design is influenced by the technical suitability, cost-effectiveness, and the availability of local skills and materials. In developing countries, the design selection is largely determined by the prevailing and proven design in the region, which in turn depends on the climatic, economic, and substrate-specific conditions (Lohri, 2012).

5.4.2.2 Biomethanation System for Organic Fraction of MSW

A biphasic design namely TEAM (TERI's Enhanced Acidification and Methanation) Process (Fig. 5.16) has been developed and tested by The Energy and Resources Institute (TERI), New Delhi (Lata et al., 2001). The plant consists of two phases–acidification and methanation. Acidification process has a total retention time of 6 days and methanation takes 1 day. Thus, total processing time is 7 days only. For acidification process, there are six reactors which are fed with substrate daily in serial order and submerged in water for microbial action. VFA rich leachate is collected from the bottom of the reactor after 6 days which is then fed into the methane reactor. The biogas produced from methane reactor is collected in an inverted drum placed over the water body. The effluent discharged from methane reactor is recycled to the acid reactor for drawing more acid rich liquor from waste.

CHAPTER SIX

WtE Best Practices and Perspectives in Africa

William H.L. Stafford
Council for Scientific and Industrial Research, Stellenbosch, South Africa
Department of Industrial Engineering, University of Stellenbosch, Stellenbosch, South Africa

6.1 MUNICIPAL SOLID WASTE MANAGEMENT AND POLICIES IN AFRICA

Suzan Oelofse
Council for Scientific and Industrial Research, Pretoria, South Africa

Globally, the management of waste is incorporated into environmental legislation in order to protect both the environment and human health from the negative effects of waste disposal (Oelofse and Godfrey, 2008). Waste management in Africa is still largely dominated by unsanitary landfilling and open dumping yet, solid waste is increasingly seen as an alternative source of renewable energy (Simelane and Mohee, 2012). The quality and availability of data in solid waste management (SWM) and generation in Africa is generally poor. This lack of access to accurate date is hampering the development of programs that will promote the efficient use of solid waste in Africa (Simelane and Mohee, 2012). The global waste sector is undergoing a paradigm shift to establish a circular economy where "waste" is viewed as "secondary resources" with the emphasis on reuse, recycling, and recovery of materials. This approach to viewing material as a renewable resource rather than waste may also provide opportunities for waste reuse and recycling in Africa (Oelofse and Godfrey, 2008). Improved waste management through reuse and recycling of waste in Africa will help the continent to achieve the Millennium Development Goal (MDG) number 7: to ensure environmental sustainability (Simelane and Mohee, 2012).

6.1.1 Status Quo of Waste Management in Africa

Africa contributes about 5% of all waste generated worldwide (Hoornweg and Bhada-Tata, 2012a,b). Population growth, urbanization, a growing middle class, and changing consumption patterns drive solid waste generation in Africa. Changes in living standards and increases in disposable incomes result in increased consumption of goods and services and consequently also an increase in the amount of waste generated (Hoornweg and Bhada-Tata, 2012a,b). Municipal solid waste (MSW) is defined as waste collected by the municipality or disposed of at the municipal waste disposal site and includes residential, industrial, institutional, commercial, municipal, and construction and demolition waste (Hoornweg et al., 2015). Data on solid waste are generally lacking and, if reported, it is mostly limited to sub-Saharan Africa (Hoornweg and Bhada-Tata, 2012a,b).

Scarlat et al. (2015) estimated waste generation in Africa using 2012 population data and waste generation data reported by Hoornweg and Bhada-Tata (2012a,b). The total MSW generated in Africa, in 2012, is estimated to be 125 million tonnes a year of which 81 million tonnes is from sub-Saharan Africa (Scarlat et al., 2015). The waste generation rate for waste in Africa is estimated to be 0.65 kg per person per day (varying between 0.09 and 3.0 kg per person per day) and is expected to increase to 0.85 kg per person per day in 2025. This translates into 169,119 tonnes of waste generated per day in 2012 and 441,840 tonnes per day in 2025. However, waste generation per capita varies considerably across countries, between cities and within cities (Hoornweg and Bhada-Tata, 2012a,b). Waste generation is generally lower in rural areas since, on average, residents are usually poorer, they purchase less products from stores and therefore, generate less packaging waste and are more likely to reuse and recycle items (Hoornweg and Bhada-Tata, 2012a,b). Africa's rapid growth in waste generation (30% between 2012 and 2025) is largely driven by urbanization and increased wealth and is not expected to stabilize before 2100 (Hoornweg et al., 2015). Urbanization in sub-Saharan Africa is expected to result in less dense cities, more akin to the United States than Japan due the availability of land. This type of cities is more likely to be associated with higher volumes of waste being generated (Hoornweg et al., 2015).

Poor governance hinders effective waste management in Africa. "Urban governance presents the most daunting and challenging task for sub-Saharan African countries in this century" (Rakodi in Lwasa and Kadilo, 2010, p. 27; Wingqvist and Slunge, 2013). In most urban areas in Africa, the municipality is responsible for providing waste services and infrastructure (UN-Habitat, 2010), but municipalities often lack the technical and financial capacity to provide efficient and effective services to all residents (McAllister, 2015). The private sector is generally better placed to provide waste services at a lower cost than municipalities (Imam et al., 2008), but only to those that are in a position to pay for the service. Other actors such as business, civil society, and the informal sector also help to strengthen the governance capacity to manage waste in Africa (Wingqvist and Slunge, 2013). In many African cities, municipalities have entered into partnerships with community-based organizations (CBOs) and the private sector to provide a more inclusive, cost-effective, and efficient waste management service (Bello et al., 2016). This approach means that municipalities are slowly shifting their functions from operations to service management (Le Courtois, 2012). The pressure to change waste management options to be more efficient and to minimize waste generation rates are to a large degree influenced negatively by the availability and capacity of existing waste management infrastructure (Hoornweg et al., 2015). In other words, there is currently little to no incentive to change to more effective and sustainable waste management options.

In low- and middle-income countries, SWM can be the single largest budget item on the municipal budget consuming 20%–50% of the annual municipal budget (Kubanza and Simatele, 2016). Between 50% and 59% of the waste management, budget can be spent on waste collection alone (Hoornweg and Bhada-Tata, 2012a,b). Therefore, being one of the most visible urban services, the effectiveness and sustainability of waste management services, can serve as an indicator for sound municipal management (Okot-Okumu, 2012). However, the success of MSW management relies heavily on social, economic, and psychological factors—including public participation, policy, and public attitudes and behavior (Ma and Hipel, 2016). Poor waste management in Africa has environmental and health concerns. Open dumping with associated uncontrolled burning is the most common waste disposal option in Africa (Hoornweg and Bhada-Tata, 2012a,b). This situation contributes to the perception that African cities are unhealthy (Simelane and Mohee, 2012). According to the UNEP (2015), 19 out of 50 of the world's biggest dumpsites are located in Africa (Fig. 6.1).

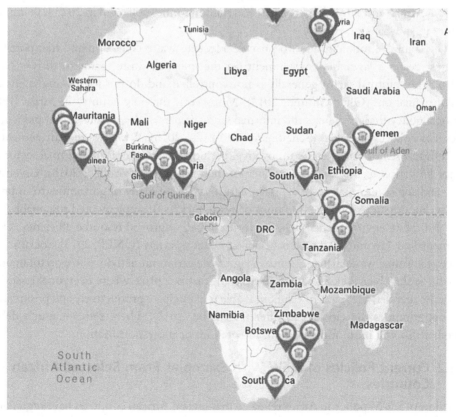

Fig. 6.1 The location of the 19 biggest dumpsites in Africa. *(Data from the Waste Atlas. http://www. atlas.d-waste.com/.)*

The waste collection rates in Africa are well below the global average waste generation rates. The collection rates vary depending on the income level of the country and by the region. High-income countries have collection rates averaging 98% while in low-income countries the average rate is 40%. In Lesotho, a low-income country, only 7% of urban households have access to waste collection services (Simelane and Mohee, 2012). The collection rates in Northern Africa are also significantly higher than in sub-Saharan Africa (Scarlat et al., 2015). The collection coverage in Africa is between 25% and 70% (UNEP, 2015). In general, the capital and major cities in Africa have waste collection systems in place. Data on waste generation and collection are generally only available at country level with some data for large cities and limited data for smaller cities (Scarlat et al., 2015). Scarlat et al. (2015) note the limitation of data uncertainties since large cities could have higher waste generation and collection rates than smaller cities. Uncertainties in the amounts of waste disposed of at landfills and the possible waste available for waste to energy (WtE) and other treatment technologies, therefore, remain problematic. Despite these challenges, Scarlat et al. (2015) estimated that 54% of the waste or 68 million tonnes were collected in Africa in 2012 (44% or 36 million tonnes in sub-Saharan Africa) and this will increase to 168 million tonnes collected in 2025 (106 million tonnes in sub-Saharan Africa).

The failure of municipalities to provide adequate waste collection and transportation systems creates a favorable environment for the informal waste sector to thrive (Noel, 2010). The informal sector generally is not controlled and do not follow any safety and health regulations (Okot-Okumu, 2012). Morocco is the only country in Africa with a national policy that recognizes the informal sector as part of the private sector and authorizes it to collect recyclables (Scheinberg and Savain, 2015). While waste management by the informal sector can often be innovative, the nonintegration of the informal sector is a major limitation to the social acceptance of their activities (Nzeadibe, 2015). Given the large and active informal waste sector in Africa, there are many opportunities to improve the livelihoods and working conditions of waste pickers. If implemented sustainably, this will also create environmental benefits, for example, improved resource efficiency, environmental quality, and the maintenance of ecosystem services (UNEP, 2013). Social acceptance of informal waste management as a legitimate economic activity is important in order to achieve the objective of an "inclusive city"—that is, a place where everyone, regardless of wealth, gender, age, race, or religion, is able to participate productively and positively in the opportunities the city has to offer (Nzeadibe, 2015). There remains much debate whether this will mean formalization, integration, or professionalization.

6.1.2 Current Policies of MSW With Examples From Selected African Countries

A review of SWM policy in Africa revealed that several African countries have regulations and policies on how waste should be managed (Bello et al., 2016). However, it appears that despite strong legislation in some countries, the overall implementation and enforcement remain weak (Makara, 2009). For example, the policy, institutional, and legal framework for the waste management sector in Egypt is not effective (National Solid Waste Management Program, NSWMP, 2011). The NSWMP (2011), therefore, set out to establish "new and effective policy, legislation and institutional arrangements for waste management at the National and Governorate level in Egypt, coupled with enhanced professional capacity, and an investment pipeline for implementation of sectoral projects at the regional and local level." The program is structured into six components, namely: institutions, policy, and legislation; investment programming and implementation; professional development; planning services and infrastructure; civil society participation; and implementation. The government of Egypt endorsed a proposal for the establishment of an Egyptian Executive Agency for Integrated Solid Waste Management (ESWA) within the Ministry of Local Development (NSWMP, 2011). The national waste management policy as envisaged by the NSWMP will be developed around the principles of self-sufficiency, waste management hierarchy, proximity, recognition of waste as a professional sector, and polluter pays principle (NSWMP, 2011).

The South African Policy on Integrated Pollution and Waste Management (DEAT, 2000) identifies integrated waste management planning, waste information systems, general waste collection, waste minimization, recycling, waste treatment and disposal, capacity building, education, and awareness as key intervention measures that are needed for efficient use and management of waste in Africa (Simelane and Mohee, 2012). In addition, policy incorporated the waste management hierarchy together with principles of the polluter pays and precautionary duty of care (DEAT, 2000).

Waste policies and legislation will at best be an exercise in futility if they are not effectively enforced and complied with (Nwufo, 2010). The mere enforcement of available legislation, including municipal by-laws, will improve the waste situation at community level in African municipalities. Illegal dumping and littering are, by default, illegal activities that should be treated as such by law enforcement officers. It is, therefore, important that enforcement officers must know what their responsibilities are under legislation, and what actions can be taken under varying circumstances (Oelofse and Godfrey, 2008). The failure of the laws and regulations for waste management is largely due to ineffective provisions and sanctions to deal with transgressors and the inability or unwillingness of officials to enforce laws (Kazungu, 2010). East African countries have policy, laws, and regulatory provisions that can limit improvements in SWM by restricting cost recovery from waste recycling and beneficiation while relying on a limited fiscal budget for waste management (Mbuligwe, 2012). In Kenya, the responsibility for waste collection, disposal, and coordination of actors involved in SWM, amongst others, are assigned to the local municipality by law, creating a barrier to private sector involvement (Bello et al., 2016).

In Ghana, the Local Government Act, 1993 (Act 462) confers power to local authorities to promulgate and enforce by-laws to regulate SWM, amongst others, but private companies cannot operate without the approval of or license from, the local authority (Van Dijk and Oduro-Kwarteng, 2007). A similar situation has also emerged in South Africa

(Oelose and Mouton, 2014). While making waste management a municipal function is seen as being crucial to ensuring that all citizens (rich and poor) receive a service, it can result in municipalities becoming gatekeepers to the waste, especially waste that can be reused, recycled, and recovered. Public-private partnerships are key to unlocking this opportunity. However, if municipalities are stuck in traditional collect-transport-dump, opportunities to move waste up the hierarchy can be lost. Currently, this problem is being somewhat bypassed in Africa as a result of a large, active informal waste sector which are able to access recyclable waste on curbside and at the landfill (UNEP, 2018).

In the recent years, with increased awareness of the environmental impacts of plastics, a number of African countries have promulgated laws to regulate the use of plastic bags. The list of countries with regulations on plastics, and the year of the regulations is summarized in Table 6.1. These regulations vary considerably from the ban of only single-use (thin) plastic bags and associated requirements for bag thickness, to complete bans on all plastic bag use.

Table 6.1 Summary of Countries That Have Introduced Regulations on Single-Use Plastic Products, or Announced Imminent Action (Sorted by Year)

Country	Year	Level	Policy	Features
Eritrea	2002	Local	Ban	Ban on the use of plastic bags in the capital Asmara
	2005	National	Ban	Ban on importation, production, sale, and distribution of plastic bags
South Africa	2004	National	Ban and levy	Ban on plastic bags $<30\,\mu m$ and levy on retailers for thicker ones
Tanzania	2006	National	Ban approved	Ban on the production, importation, sale and use of plastic bags $<30\,\mu m$
Zanzibar	2006	National	Ban	Ban on the importation, distribution, and sale of plastic bags $<30\,\mu m$
Botswana	2007	National	Levy	Voluntary levy on plastic bags; retailers decided if and how much to charge
	2017	National	Ban announced	Government is considering the introduction of a ban on plastic bags $<24\,\mu m$
Uganda	2007	National	Ban	Ban on lightweight plastic bags $<30\,\mu m$
Rwanda	2008	National	Ban	Total ban on the production, use, importation, and sale of all polyethylene bags
Egypt	2009	Local	Ban	Total ban on the use of plastic bags in Hurghada
Morocco	2009	National	Ban	Ban on the production, importation, sale, and distribution of black plastic bags
	2016	National	Ban	Ban on the production, importation, sale and distribution of plastic bags
Chad	2010	Local	Ban	Total ban on the importation, sale, and use of plastic bags in the capital city, N'Djamena
Zimbabwe	2010	National	Ban and levy	Ban on plastic bags $<30\,\mu m$ and levy on thicker ones
	2017	National	Ban	Total ban on Styrofoam products

Table 6.1 Summary of Countries That Have Introduced Regulations on Single-Use Plastic Products, or Announced Imminent Action (Sorted by Year)—cont'd

Country	Year	Level	Policy	Features
Republic of the Congo	2011	National	Ban announced	The government announced a ban on the production, importation, sale, and use of plastic bags, but did not announce when it would take effect
Mali	2012	National	Ban	Ban on the production, importation, possession, sale, and use of nonbiodegradable plastic bags
Mauritania	2013	National	Ban	Ban on the manufacture, use, and importation of plastic bags
Niger	2013	National	Ban	Ban on production, importation, usage, and stocking of low–density smooth plastic, and packaging bags
Cameroon	2014	National	Ban	Total ban on nonbiodegradable plastic bags
Côte d'Ivoire	2014	National	Ban	Ban on the importation, production, use, and sale of nonbiodegradable plastic bags <50 μm
Burkina Faso	2015	National	Ban	Ban on production, import, marketing and distribution of nonbiodegradable plastic bags
Gambia	2015	National	Ban	Ban on the use and importation of plastic bags
Malawi	2015	National	Ban	Total ban on the use, sale, production, exportation, and importation of plastic bags <60 μm
Somalia	2015	Local	Ban	Total ban on plastic bags in Somaliland
Ethiopia	2016	National	Ban	Ban on production and importation of nonbiodegradable plastic bags <30 μm
Guinea-Bissau	2016	National	Ban	Total ban on the use of plastic bags
Mauritius	2016	National	Ban	Ban on the importation, manufacture, sale, or supply of plastic bags
Mozambique	2016	National	Ban	Ban on the production, importation, possession, and use of plastic bags <30 μm
Senegal	2016	National	Ban	Ban on the production, importation, possession, and use of plastic bags <30 μm
Benin	2017	National	Ban approved	Total ban on import, production, sale, and use of nonbiodegradable plastic bags
Cape Verde	2017	National	Ban	Total ban on the sale and use of plastic bags
Kenya	2017	National	Ban	Total ban on the importation, production, sale, and use of plastic bags
Tunisia	2017	Public-private agreement	Ban and levy	Ban on the production, importation, and distribution in large supermarkets of plastic bags <30 μm and levy on consumers for thicker ones. The action does not affect smaller shops and street food markets

Data source: UN Environment International Environmental Technology Centre (IETC) research and record, 2018

Modified from UNEP, 2018. Africa Waste Management Outlook. United Nations Environment Program.

There are notable gaps in policy, "piecemeal" regulations, and nondomestication of international agreements have been identified as weak links, making Africa vulnerable for illegal dumping of hazardous waste from the developed countries (Osibanjo, 2002). African countries are significant players in the negotiations resulting the formulation of environmental treaties, but there are numerous barriers to implementation once these agreements have been signed (Osibanjo, 2002). The need to integrate environmental conventions into domestic policies and legislation is only beginning to be acknowledged (Gray, 2003).

In summary, waste management in Africa is failing because of an unsupportive governance environment to deal with the increasing waste volumes, increased urbanization, and economic growth. The most common waste management option employed in Africa remain open dumping and uncontrolled burning of waste. Where waste policies and legislation are in place, implementation and enforcement lag behind, resulting in ineffective waste management. The private sector is often better positioned to provide efficient and effective waste management services, but policies are generally not supportive of partnerships. There is thus a need for policies that are supportive of partnerships to improve waste management services, but also more harmonized policies between African countries to allow for the development of regional waste management plans toward implementing a secondary resources economy. The transboundary movement of waste in Africa needs to be controlled through the domestication of conventions and treaties to avoid Africa being as easy target for illegal dumping. However, the controlled movement of waste and secondary resources between African countries should be supported to ensure economies of scale for recycling and alternative waste treatment options. This approach will facilitate the development of local and regional processing capacity for secondary resources in Africa and encourage responsible waste management, and creating job opportunities in waste and the circular economy on the African continent.

6.2 ENERGY ACCESS IN AFRICA: CURRENT SITUATION AND DIFFICULTIES TO INCREASE ACCESS

Max Mapako and William H.L. Stafford

Council for Scientific and Industrial Research, Pretoria, South Africa
Council for Scientific and Industrial Research, Stellenbosch, South Africa; Department of Industrial Engineering, University of Stellenbosch, Stellenbosch, South Africa

Department of Industrial Engineering, University of Stellenbosch, Green Economy Solutions, Natural Resources and the Environment Unit, Council for Scientific and Industrial Research, Stellenbosch, South Africa

Energy interacts with society, the economy, and the environment in several ways and energy is a ubiquitous metatechnology. Energy is a domestic necessity and a factor of production; enabling a variety of services such as transportation, heating, and food production. The price of energy is a significant cost that directly affects the price of other goods and services (NEPAD, 2000). The access to secure, sustainable, and affordable energy is also seen as being essential for achieving the MDGs, such as the reduction in hunger and poverty, improving education and communication, enhancing health-care services, and responding to climate change (IIASA, 2005).

Energy sufficiency and security is a key to development and prosperity since it is an essential domestic necessity and required for the total factors of production and the public services that improve the quality of life. The inability of many African countries to provide good and adequate energy services has been a major constraint to their development. Energy is a nexus for sustainable development since it a domestic necessity and a factor of production. While the need to expand energy access is a widely recognized developmental goal, there is no clear and universally accepted definition of energy. The International Energy Agency (IEA) (2011a, p. 12) defines Universal Modern Energy Access as "a household having reliable and affordable access to clean cooking facilities, a first connection to electricity and then an increasing level of electricity consumption over time to reach the regional average." Energy access, therefore, refers to a progression up an energy ladder with increasing energy needs—from energy for basic human needs, productive uses, and modern societal needs (see Fig. 6.2).

Only one-third of the African population has access to electricity and Africa has been termed "the dark continent" based on this fact, which can be observed from satellite imagery (see Fig. 6.3). The primary energy use per capita in Africa is about 11 and 5 times

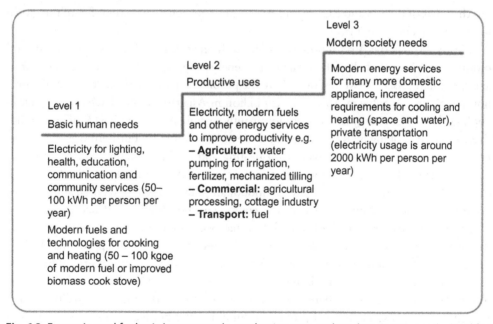

Fig. 6.2 Energy is used for basic human needs, productive uses, and modern society needs. *(Modified from AGECC—The UN Secretary-General's Advisory Group on Energy and Climate Change, 2010. Energy for a Sustainable Future. Report and Recommendations. Available from: https://www.un.org/chinese/millenniumgoals/pdf/AGECCsummaryreport%5B1%5D.pdf.)*

Fig. 6.3 The Earth at night from NASA satellite imagery. *(From https://www.nasa.gov/topics/earth/earthday/gall_earth_night.html.)*

less than that of the US and EU-27 and Africa and represents only about 5% of the global total primary energy demand.

Nearly, 80% of the African population depends on traditional biomass for cooking and biomass contributes to half of the total primary energy supply of Africa. Sub-Saharan Africa and developing Asia are two regions where the problem is particularly acute; with 653 million people in sub-Saharan Africa and 1.9 billion in Asia that do not have access to clean cooking fuels. The energy access is predominantly a rural problem with 81% of households that lack clean cooking fuels residing in rural areas (IEA, 2011a,b). Furthermore, 90% of the energy needs of the poor originate from heating and cooking energy demands whereas electricity is used for lighting and entertainment needs (World Bank, 2008). The use of firewood for cooking is a major source of indoor air pollution where women and children are more exposed and burdened with the consequent health effects (Bruce et al., 2002). The prolonged exposure to indoor pollution has been implicated in the increasing incidence of respiratory diseases in developing countries. It accounts for approximately 10% of disease-related deaths in Africa (Smith, 2000) and, with the current patterns of firewood usage, will result in nearly 10 million premature deaths by 2030 because of indoor air pollution (Bailis et al., 2005). There are several other social and gender inequality impacts related to energy use in Africa. In rural communities, the burden of household activities such as firewood collection falls primarily on children and women, who gather firewood on foot is often walking long distances with heavy loads. The loads can be 20 kg and the distances up to 10-km round-trip (IEA, 2006; McPeak, 2003). Since many women in rural communities spent considerable time collecting firewood, they sacrifice valuable opportunities in terms of education or other income generating activities. In addition, firewood and charcoal are biomass resources that being used in a nonrenewable manner and are contributing to deforestation in Africa (FAO, 2016).

Wood fuels are an energy resource that are widely available and, when utilized sustainably, can have a major role in the worldwide transition to renewable energy both in the developing and in the developed countries. However, the current trend indicates forest resources are diminishing because of overexploitation due to deforestation for timber, wood fuels, charcoal, and agriculture. To place this in context, it is estimated that persons in East Africa (Nairobi, Kenya) that rely upon charcoal for cooking consume 242 kg of charcoal annually and this requires 1500–3500 kg of wood using the traditional charcoal production processes. Since the mean annual increment of forest growth is approximately $2.35 \, m^3/ha/year$, but the estimated annual removal rate is $6.4 \, m^3/ha/year$ the harvesting of wood for charcoal is clearly not renewable and there is a consequential per capita deforestation footprint of approximately 1 ha per year (Kammen, 2006; Malimbwi et al., 2010). Between 1850 and 1998, the conversion of forests to other land use contributed approximately 136 billion tonnes of carbon to the atmosphere which represents approximately one-third of the total emissions during that period (Malmsheimer et al., 2008). Since approximately 90% of world's population growth is taking place in the developing countries and their populations will be nearly 10 times larger than that of the developed countries in two generations (WEF, 2018), current trends suggest an accelerated use of forest resources to meet the growing energy demands. It is, therefore, essential to ensure that there are improvements in land management and biomass resource usage with wastes used for energy and other valuable products.

The household energy consumption of the poor is primary for cooking needs (Moulot, 2005; Fall, 2010) and is typically met by using traditional biomass cook stoves with efficiencies <20% (Sanga and Jannuzzi, 2005). The patterns of energy usage in societal development have been expressed in terms of an energy ladder; with the least efficient and most polluting fuels at the bottom of the more efficient use polluting fuels at the top. The lowest rung of the ladder consists of households utilizing wood fuels such as firewood and animal wastes. At an intermediate rung on the energy ladder households have shifted toward cleaner burning fuels, but still with notable emissions (i.e., charcoal and paraffin). At the highest rung of the ladder households use clean household fuels such as electricity, liquefied petroleum gas (LPG), and the biofuels. This energy ladder tends to be an oversimplification since households do not perfectly substitute one fuel for another often utilize different options alongside one another so a more appropriate conceptual representation is the energy stack (Masera et al., 2000; Pachauri and Spreng, 2003; Daurella and Foster, 2009). The complexity of switching energy sources depends on underlying economic social and environmental factors and has been approached from several angles—both the supply side switching to more modern cooking fuels (paraffin, LPG, and biofuels) and improved stove efficiency (Goldenberg, 2000). However, there are other factors aside from income that contribute to household fuel switching, including increased scarcity of firewood, availability of other fuel options, urbanization, and human settlement patterns (FAO, 1993; Dewees, 1989). Urbanization typically involves a switch from wood to charcoal, especially since there is often a small difference in price of firewood and charcoal in urban centers (Van der Plas and Abdel-Hamid, 2005) and charcoal is a cleaner burning with improved storage and transportation (Van der Plas and Abdel-Hamid, 2005; Arnold et al., 2003).

There are also huge inequalities and disparities in energy services in Africa where modern and traditional energy systems often coexist. Over 80% of the electricity generated in the continent is from fossil fuels (IEA, 2011b) and this contributes to the increasing costs of electricity generation because of the volatility of fossil fuel prices. There has been a lack of investment in energy and countries of sub-Saharan Africa spend <3% of their gross domestic product (GDP) on their power sector, with operating costs absorbing three quarters of the total spending (Eberhard et al., 2008); that leaves 0.5% of GDP for investment in the energy sector (Sokona et al., 2013). The power sector in sub-Saharan Africa faces huge challenges that including low generation capacity, high costs of electricity generation, high transmission loses, unstable and unreliable electricity supply, and low electrification rates. The installed generation capacity per million people in sub-Saharan Africa is the lowest in the World—with a generation capacity of 92.3 MW per million people and a large contribution of this is from South Africa, which accounted for 51.8 MW per million people of the sub-Saharan Africa total (IEA, 2011b). There has also been a slow growth in electricity generation capacity in Africa. While generation capacity grew at an average annual rate of 5%–9% in the last 20 years in the developing nations of Asia and the Pacific, the generation capacity in sub-Saharan Africa grew <2% over the same period [Energy Information Administration (EIA), 2011]. Electricity transmission and distribution losses are also high in many African countries and there is grid instability leading to insecurity of electricity supply. For example, electricity distribution losses in several African countries are over 20% (EIA, 2011) and many countries experience between 50 and 170 power outages per year. Since an average power outage can last between 5 and 12 h, there are significant socioeconomic costs that include damage to equipment, forgone sales, and health risks from food spoilage (Ramachandran, 2008; Eberhard et al., 2008). Many businesses in Africa rely on their own diesel-powered generators as a backup to the national grid, which adds costs and increases reliance on fossil fuels.

Against this backdrop are growing African populations, increasing urbanization and material consumption that are leading to increasing wastes being generated and concentrating in the cities; with the resultant growing challenges of waste management. It is estimated that less than half of waste is currently being formally collected for central processing and safe disposal in Africa (UNECA, 2009). WtE, therefore, represents a synergy and opportunity to ensure wastes are safely treated while providing energy for sustainable development.

6.3 WtE TO INCREASE ENERGY ACCESS IN AFRICA

William H.L. Stafford

Council for Scientific and Industrial Research, Stellenbosch, South Africa
Department of Industrial Engineering, University of Stellenbosch, Stellenbosch, South Africa

The world's urban population is rapidly increasing and >90% of future population growth is expected to occur in the large cities in the developing countries. In the developing world, Africa has experienced the highest urban growth during the last two decades at 3.5% per year and this trend is expected to continue until 2050 (AfDb, 2011). The consequences of urbanization and associated consumerism with increased material and energy consumption led to a concentration of wastes that requires safe and effective disposal methods. As towns and cities grow, there is a need to find new landfill sites as current landfills reach their capacity, and the real estate value of the land reaches a premium. Many of the existing municipal landfill sites in the large urban centers are reaching their capacity after decades of landfilling, and alternative options for disposal are being sought. Several recent projects are reducing waste going to landfill through increased recycling and recovery of valuable materials and energy from current wastes. In addition, several landfill sites that contain legacy wastes, often from decades of landfilling, have been capped and methane-rich gas harvested to mitigate the environmental impacts greenhouse-gas emissions and generate electricity. There are some more recent developments where municipal organic wastes from households and sewage are being used to produce biogas to provide heat and electricity either for powering waste treatment or sold to provide revenue for the waste management company.

The provision of energy from wastes in urban centers is an obvious synergy due to the concentration of large amounts of waste not only from both household and industries. The value gained from WtE is far greater than the value of the energy obtained, since the processing of waste with value adding helps to prevent and mitigate many of the impacts to the environment and human health that would have occurred without adequate waste collection, and effective processing and disposal.

6.4 WtE EXPERIENCES FROM AFRICA: CASE STUDIES

William H.L. Stafford

Council for Scientific and Industrial Research, Stellenbosch, South Africa
Department of Industrial Engineering, University of Stellenbosch, Stellenbosch, South Africa

In most urban areas in Africa, the municipality is responsible for providing waste services and infrastructure (UN-Habitat, 2010), but municipalities often lack the technical and financial capacity to provide efficient and effective services to all residents (McAllister, 2015). The private sector is generally better placed to provide waste services at a lower cost than municipalities (Imam et al., 2008), but only to those that are in a position to pay for the service (Wingqvist and Slunge, 2013). In many African cities, municipalities have entered into partnerships with CBOs and the private sector to provide a more inclusive, cost-effective, and efficient waste management service (Bello et al., 2016). This approach means that municipalities are slowly shifting their functions from operations to service management (Le Courtois, 2012). However, there are little incentives to minimize waste generation and achieve more sustainable waste management (Hoornweg et al., 2015).

In low- and middle-income countries, SWM budget can consist of 20%–50% the total (Kubanza and Simatele, 2016), with over half of the waste management budget spent on only on the collection of waste (Hoornweg and Bhada-Tata, 2012a,b). The effectiveness and sustainability of waste management services can serve, therefore, serve as an indicator of sound municipal management (Okot-Okumu, 2012). Poor waste management in Africa has environmental and health concerns. Open dumping with associated uncontrolled burning is the most common waste disposal option in Africa (Hoornweg and Bhada-Tata, 2012a,b).

The failure of municipalities to provide adequate waste collection and transportation systems creates a favorable environment for the informal waste sector to thrive (Noel, 2010). The informal sector generally is not controlled and do not follow any safety and health regulations (Okot-Okumu, 2012). While waste management by the informal sector can often be innovative, the nonintegration of the informal sector is a major limitation to the social acceptance of their activities (Nzeadibe, 2015). Given the large and active informal waste sector in Africa, there are many opportunities to improve the livelihoods and working conditions of waste pickers. If implemented sustainably, this will also create environmental benefits, for example, improved resource efficiency, environmental quality, and the maintenance of ecosystem services (UNEP, 2013).

6.4.1 WtE Projects in South Africa

William H.L. Stafford

Council for Scientific and Industrial Research, Stellenbosch, South Africa
Department of Industrial Engineering, University of Stellenbosch, Stellenbosch, South Africa

South Africa this has typically applied in the pipe landfilling solution to the disposal of wastes. Landfilling is seen as a cheap solution and municipalities offering lower gate fees except in cities where land space is becoming a premium. It is estimated that 65% of the waste approximately 38 million tonnes could be diverted from landfill with materials recycled and energy recovered, but only about 10% of materials are currently being recycled or energy being recovered. This represents a significant loss of valuable resources that could otherwise be recovered and recycled; in addition to the landfilling of waste that creates significant financial, socioeconomic, and environmental burdens (Nahman, 2011). This is being addressed by the National Environmental Management: Waste Act (NEM:BA, No. 59 of 2008) that calls for increased diversion of waste away from landfill toward reuse, recycling, and recovery. Further, the National Waste Management Strategy (Fig. 6.4) provides a guide for the appropriate management of waste through establishing a waste management hierarchy that calls for the avoidance and minimization of waste production and not just end of pipe recovery, recycling, and treatment solutions.

Countrywide, an estimated 20 million tonnes of MSW is generated per annum, of which about 25% consists of mainline recyclables (paper, plastics, glass, tins, and tires). An estimated 3.39 million tonnes of packaging was consumed in South Africa, of which only 52.6% was recycled [Department of Environmental Affairs (DEA) 2012; Packaging SA, 2015], with the remainder disposed of at landfills. The organic waste of South Africa's six largest city municipalities has potential to produce 207 MW electricity from the organic wastes (SACN, 2014).

Currently, South Africa's largest city, Johannesburg, produces between 1.3 and 2.6 million tonnes/annum (with a median value of 1.6 million tonnes) of municipal waste. This includes domestic and industrial wastes. Based on the data collected over a period of 3 years (2007/08–2009/10), the weekly waste collection from households contributes the highest percentage of the waste stream at 54.7% or 646,000 tonnes per annum. The remaining 45.3% comprises other sources such as street cleaning (6.3%), garden refuse (9.7%), builders' rubble in its different categories (12.2%), and waste cleared from illegal dumping areas contributing 16.5%. This makes illegal dumping the second biggest waste stream illustrating that illegal dumping is still a serious problem within the City of Johannesburg.

Fig. 6.4 Waste hierarchy from the National Waste Management Strategy, 2011. *(Modified from National Waste Management Strategy, 2011. https://www.environment.gov.za/sites/default/files/docs/national waste_management_strategy.pdf.)*

Table 6.2 Current and Future Plans for the Installation of Landfill Gas Power Plants (2015–18) in the City of Johannesburg at the Main Municipal Solid Waste Dump Sites (Ener-G Submission to NERSA)

Landfill Site in the City of Johannesburg	Maximum Electricity Generation Potential (MW)
Lindro Park	2
Marie Louise	5
Ennerdale	1
Goukoppies	4
Robinson Deep	6

Data from http://www.nersa.org.za/Admin/Document/Editor/file/Consultations/Electricity/Presentations/Ener%C2%B7G%20Systems%20(Pty)%20LTD.pdf.

Approximately, 15%–35% of the waste is organic and termed the organic fraction of municipal solid waste (OFMSW). The OFMSW is a component of the wastes disposed of to landfill and is, therefore, estimated to be 244,896–571,424 tonnes per annum, and with energy value 9 MJ/kg, is at least 2204 TJ per annum is theoretically available. However, only 35%–40% of this is likely to be available to anaerobic digestion to yield methane—this yields an estimated 841 TJ per annum. Most of the OFMSW is collected by Pikitup, the City of Johannesburg official waste management service provider; who services the entire 1625 km^2 that is Johannesburg, collecting and disposing of 1.4 million tonnes of domestic waste generated by the city's 3.2 million citizens every year, and 17,000 businesses. The company owns 12 waste management depots strategically located throughout the city, 42 garden refuse sites, and five landfill sites and one incinerator that was located at Robinson Deep landfill site which has since been closed. All its landfill sites comply with permit requirements and are licensed by the National Department of Water Affairs and Forestry. The combined total available airspace and life spans for all four operating sites of the City of Johannesburg is currently 8 years with Robinson Deep contributing about 17 years, Goudkoppies having 6 years while Ennerdale and Marie Louise both only have a year left. The City of Johannesburg, therefore, has a critical need for alternative disposal options. There are current plans to use the existing landfills that have accumulated much of this waste, and to cap them and harvest the biogas to generate electricity and feed into the national grid (Ener-G[1]) with development Robinson Deep and Marie Louise landfills (Johannesburg) (Table 6.2).

Similar examples of successful landfill gas WtE are found in other South African cities; such as in Durban with two landfill gas-to-electricity projects at Marianhill (1 MW) and Bissasar Rd. (6.5 MW).

[1] http://www.localgovernment.co.za/metropolitans/view/2/city-of-johannesburg-metropolitan-municipality#service-delivery.

There is also an opportunity for generating energy from sewage and the municipal wastewater treatment works in the City of Johannesburg are similarly being upgraded in order to capture the energy from the sewage, while also assisting with wastewater treatment. In the City of Johannesburg, it is estimated that 87% of sewage is captured since this is the number of households connected to the waterborne sewage system.[2] The energy that can be generated from anaerobic digestion of the sewage can provide biogas to fulfill a large portion of the wastewater treatment with heat and electricity. This has recently implemented at the Northern Water Works WWTP and others in Johannesburg and elsewhere in South Africa are proposed to follow.[3]

In the City of Cape Town, the biogas WtE plant in Athlone Industria sources was commissioned in 2017 by New Horizons energy (Fig. 6.5). Approximately 10% of Cape Town's MSW (ca. 8000 tonnes per day) is sorted and the organic waste fraction separated for biogas production using anaerobic digestion. The WtE plant receives about 643 tonnes of waste a day and through recycling and organic waste extraction, achieves an 88% diversion rate from landfill. Approximately 670 m^3 per day of biomethane is generated; which amounts to 600 GJ per year and is equivalent to a 5 MW biogas-to-electricity production.[4] However, the biomethane is not used to generate electricity; but cleaned, upgraded and pressurized, and sold to Afrox as a natural gas substitute [bio-compressed natural gas (Bio-CNG)]. In addition, during the biogas cleanup, the carbon dioxide that is removed is also further purified and pressurized and sold to the beverage industry for carbonation.

There are several other notable smaller-scale examples of WtE in South Africa. For example, Agama and TradePlusAid have installed a number of small (1 kW) domestic biogas digesters with gas used for cooking. BiogasSA and Cape Advance engineering have installed biogas plants of several hundred kW energy production capacity at Morgan Abattoir (Springs) and Humphries Boerdery (Bela-Bela), respectively; and are used to treat animal manures and abattoir waste while generating biogas for heat and electricity.

The industrial-scale biogas plant in Bronkhorspruit was developed by Bio2Watt and the engineering carried out by Bosch. It has an electrical production capacity of 4.4 MW with a purchase power agreement for sale of electricity to the BMW car production plant in Roslin. The electricity from biogas will cover approximately a quarter of electricity needs of the BMW plant over the 10-year duration of the power purchase agreement. The local municipality, Tshwane, facilitated this business arrangement by providing access to the municipal grid in order to deliver electricity to the end user. The majority of feedstock for the biogas plant is generated by an adjacent farm that has a feedlot for approximately 20,000 head of cattle and 12,000 sheep. The manures and slaughterhouse waste from the feedlot is supplemented from regional municipal organic waste sourced from food-processing industries.

[2] http://pods.dasnr.okstate.edu/docushare/dsweb/Get/Document-8544/BAE-1762web.pdf.
[3] http://www.cityenergy.org.za/uploads/resource_171.pdf.
[4] Assumes 25% efficiency in the conversion of biogas to electricity.

Fig. 6.5 The New Horizons Energy biogas plant in the City of Cape Town. The plant receives about 643 tonnes of municipal waste per day and generates upgraded biogas from the organic fraction of municipal. Wastes using anaerobic digestion. Top: waste receiving at the plant. Bottom: The Anaerobic digesters. *(Photo by William H. L. Stafford.)*

6.4.2 WtE Project in Zimbabwe

Madzore Mapako

Council for Scientific and Industrial Research, Pretoria, South Africa

In Zimbabwe, urban local authorities are responsible for urban SWM. The current waste management practice in Zimbabwe is the traditional closed system. The system involves the stages of waste generation, storage, collection, transfer, transport, processing, and disposal. It primarily focuses on the disposal of solid waste without controlling its generation and promoting reduction, reuse, recycle, and recovery. Managed solid waste disposal sites (SWDS) are mainly found in urban centers while unmanaged SWDS, which are more like those in rural areas are found in growth points (Feresu, 2013) and other small settlements.

The disposal of waste at landfills is the most common SWM practice in Zimbabwe (ZimStats, 2016) (Fig. 6.6). Most of the landfills do not meet the criteria set out by the Environmental Management Agency (EMA) for engineered landfills, making it difficult to recover energy from sites especially in municipalities, towns, and councils.

The treatment and disposal of municipal, industrial, and other solid waste releases large amounts of methane (IPCC, 1996), and in 2006 MSW generated 25.45 Gg of methane (MEWC, 2016) that was not captured.

The Constitution of the Republic of Zimbabwe accords every person the right to a clean and safe environment that is not harmful to his or her health and well-being according. Zimbabwe's environmental legislation is comprehensive. SWM in Zimbabwe is regulated through several ministries:

- Environment, Water and Climate (MoEWC) through the Environmental Management Act, Chapter 20:207.

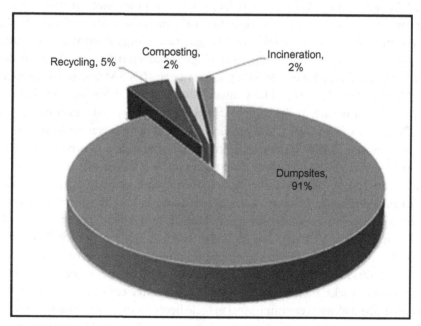

Fig. 6.6 Waste disposal methods. *(Data from MEWC, 2014. Zimbabwe Integrated Solid Waste Management Plan. Ministry of Environment, Water and Climate, Government of the Republic of Zimbabwe, Harare.)*

- Local Government Public Works and National Housing (MLGPWNH) through the Urban Councils Act, Chapter 29:15 and the Rural Councils Act, Chapter 29:13.
- Health and Child Care (MoHCC) through the Public Health Act, Chapter 15:09.

The most important piece of legislation is the Environmental Management Act Chapter 20:27 which governs waste management, with Statutory Instrument (SI) 6 of 2007 regulating the disposal of solid waste and effluent, based on the polluter pays principle. SI 10 of 2007 of the same Act provides for the licensing for generation, storage, use, recycling, treatment, transportation, or disposal of hazardous waste while SI 12 of 2007 regulates storage and management of herbicides, pesticides, fungicides, or any toxic substances. Plastics and plastic packaging material is regulated through SI 98 of 2010.

The National Environmental Policy and Strategies 2009 include the right to a clean environment; the polluter pays principle and the precautionary principle with regard to anticipated negative impacts of solid waste on the environment. The National Climate Change Response Strategy (NCCRS) (MEWC, 2015a,b) and the Integrated Solid Waste Management Plan (ISWMP), which is based on the waste hierarchy of reduce, reuse, recycle and dispose, spells out strategies, and implementation plans for management of waste in Zimbabwe.

Other pieces of legislation, which have a bearing on waste management, include the National Energy Policy. One of its objectives is the promotion of use of renewable energy sources to complement the conventional sources of energy (GoZ, 2013). This has driven the current wave of initiatives to generate energy from waste. Already agricultural and wood waste are being used to generate electricity at Triangle Limited (45 MW), Hippo Valley Estate (38 MW), Green Fuels, (18.3 MW), and at Border Timbers (0.5 MW) (ZERA, 2016). The Climate Policy in its mitigation and adoption and development of low carbon development pathway, calls for reduction of waste sector methane emissions from SWD, and promotion of energy generation from waste.

The collection of MSW is the responsibility of both urban and rural local authorities, with very few private organizations doing the same on behalf of the local authorities. From residential properties, waste is collected once a week, one bag or bin per household. Industrial and commercial properties waste is collected daily. The collection system being practiced is a combination of curbside collection and block (communal) collection (Muswere and RodicWiersma, 2004). Waste collection is a major challenge faced by urban local authorities. However, the collection rate has been improving over the years from a low of 30% in 2006 (MEWC, 2015b) to 81% in 2014, despite challenges of proper refuse collection trucks, their maintenance and an effective collection system. Most local authorities in Zimbabwe were using aged vehicle fleets for their waste management operations. The capacity in terms of equipment was generally below 20%, with tractors dominating across the surveyed local authorities at 40%, with little proper waste management equipment such as compactors (Feresu, 2013).

Weight of the solid waste is estimated through volume to mass conversions, as most local authorities do not have weighbridges at the disposal sites. Results from research work on SWM in Zimbabwean indicate the need to have an integrated SWM system (Muswere and RodicWiersma, 2004; Tsiko and Togarepi, 2012; Phiri and Mwanza, 2013; TARSC, 2010; Ngwenya and Jonsson, 2001). MSW mainly includes food waste, garden waste, paper, cardboard, wood, textiles, rubber, leather, plastics, metal, and glass (pottery and china), amongst many others (IPCC, 1996). The collection of MSW has been increasing, with households having the largest contribution to SW disposed in Zimbabwe (ZimStats, 2016), as shown in Fig. 6.7.

The proportion of urban population with access to municipal waste collection in Zimbabwe has increased from 21% in 2012 to 76% in 2015 and amounted to 1339 metric tonnes in 2015 (MEWC, 2014; ZimStat, 2016). The characterization of MSW in Zimbabwe remains a challenge for most urban local authorities. As reported in the Service Level Benchmark (SLB) report, there is no meaningful waste characterization done by local authorities. The University of Zimbabwe's Environmental Sciences Institute sampled MSW from 11 urban local authorities and yielded the composition given in Fig. 6.8.

Harare City Council in its quest to generate energy from waste deposited at its dumpsites carried out a characterization exercise and came up with waste stream

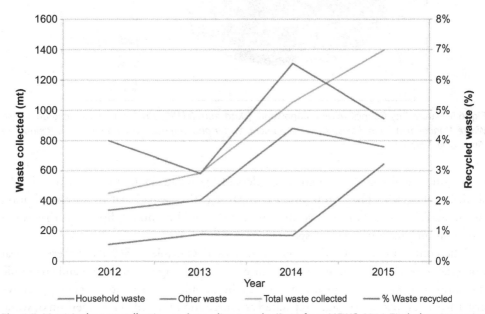

Fig. 6.7 Municipal waste collection and recycling trends. *(Data from MEWC, 2014. Zimbabwe Integrated Solid Waste Management Plan. Ministry of Environment, Water and Climate, Government of the Republic of Zimbabwe, Harare.)*

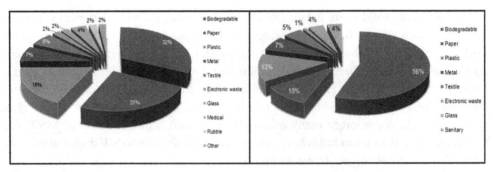

Fig. 6.8 Comparison of national (left) and urban (right) MSW composition in Zimbabwe. *(Data from MEWC, 2014. Zimbabwe Integrated Solid Waste Management Plan. Ministry of Environment, Water and Climate, Government of the Republic of Zimbabwe, Harare.)*

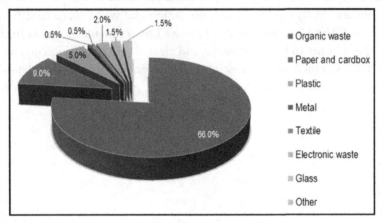

Fig. 6.9 Harare City Council MSW composition. *(Data from MEWC, 2014. Zimbabwe Integrated Solid Waste Management Plan. Ministry of Environment, Water and Climate, Government of the Republic of Zimbabwe, Harare.)*

composition which is dominated by biodegradable waste (Harare City, 2016). This is shown in Fig. 6.9. The composition compares with the national mean composition for urban local authorities shown in Fig. 6.8. It should be noted, however, that waste composition varies in time and space.

The composition of MSW is significantly biodegradable (>56%) and there is great potential to reduce the amount send to disposal sites though composting and anaerobic digestion. The waste that is finally disposed at the landfills can be treated for energy recovery. Moreover, plastics could be separated at source and recycled. This diversion of waste for material and resource recovery especially energy could signal a downturn on emissions from waste and improved waste management practices from the generated funds.

Cognizant of the challenges to collect, transport, and dispose solid waste, most urban local authorities have embarked on initiatives to manage waste through generation of energy. Bulawayo, Kwekwe, Mutare, and Harare all have plans to generate energy from either solid waste or wastewater treatment plants. Some of the projects are at implementation stage (e.g., Harare biodigester), while the majority are at tendering of prefeasibility stage; for example, the Pomona dumpsite. Several other projects are in progress or being planned:

i. Harare City Council Mbare Waste to Energy Project: The project is funded by the European Union (EU). The project is located at the back of Matapi flats near the Mukuvisi river. The Mbare digester system will have a total size of 800 m^3 (4 × 200 m^3 digesters) and will be co-fed with sewage and organic agricultural waste from the Mbare market place. The biogas is expected to run a 100–200 kVA generator for power generation and the electricity will be used in Mbare area. Slurry from the digesters will be air-dried in drying beds for use as organic fertilizer.

ii. Firle Biogas Project: The Ministry of Energy and Power Development is encouraging municipalities with biogas digesters at their wastewater treatment plants to harness biogas for electricity generation, and those without to adopt the biogas concept. As a result, the City of Harare has contracted a company, Energy Resources Africa (ERA) to refurbish the existing biogas digesters and install a generating unit at its Firle Sewage Treatment Plant to generate 2.5 MW of electricity for use at the WWTW. This will reduce pressure on the grid. The biogas generators are already (end of 2017) in Zimbabwe awaiting installation.

iii. Mutare City Council. Mutare, the fourth largest city in Zimbabwe, may soon see a new WtE plant. Working in partnership with the Harare Institute of Technology's Climate Change Research Centre and Astra Innovations, a German technological firm, a local entrepreneur has signed a Memorandum of Understanding with the City of Mutare to convert sewage at the Sakubva and Yeovil wastewater treatment plant. The project will convert sewage sludge through a process of advanced thermal distillation into diesel, natural gas, electricity, and char.

iv. Bulawayo City Council (BCC). The BCC has granted a United Kingdom-registered firm—Pragma Leaf Consulting—permission to establish a $68 million WtE plant with a potential to create up to 2000 jobs. The proposed plant is expected to produce 110,000 L of biodiesel and 2.2 MW of electricity, according to the latest council minutes. The council showed that Pragma had completed the study and results had indicated that waste generated in Bulawayo was sufficient for the establishment of a WtE plant. "They therefore, intent to establish a $68 million waste to energy plant, producing 110,000 L of biodiesel and 2.2 MW of electricity and creating an employment for 120 plant operatives and further jobs in downstream industries," the council said. The agreement conditions stipulated that Pragma would bring additional refuse removal compactors to help improve waste collections and their operations would not use portable. The conditions also indicate that only 5% of waste would be delivered to the landfill thus prolonging its lifespan.

6.4.3 WtE Project in Kenya

Ben Muok

Centre for Research, Innovation and Technology, Jaramogi Odinga University, Nairobi, Kenya, East Africa

With the fast-growing urbanization in Kenya, waste generation in cities is set to increase and management will continue to pose a great challenge. The urban population has increased from 5.4 million in 1999 to 12.5 million in 2009 and many cities, such as Mombasa, Kisumu, Nakuru, and Eldoret, have witnessed an exponential increase in the generation of solid waste. By 2030, it is projected that about 50% of the Kenyan population will be urban residents (NCPD, 2013). It is estimated that the current waste generation is 4 million tonnes per year, and predicted to double by 2030 as the urban population grows.

This rise has not necessarily been followed by an increase in the capacity of the relevant urban authorities to deal with the challenge of SWM. In Nairobi, for instance, about half of the 1500 tonnes/day solid waste produced is not being collected (Haregu et al., 2016) while in Kisumu only 20% is effectively collected. The collected waste is disposed at a dumpsite within the city while the rest of the wastes end up in the backstreets, markets, roadsides, and open spaces—this is particularly true in the informal urban settlements that lack municipal services. Kenya lacks a comprehensive response to SWM and in most urban areas where waste is collected it is done so in a haphazard and unregulated manner. The Dandora dumpsite in Nairobi and the Kachok dumpsite that serves Kisumu cities are rapidly reaching capacity and there is an initiative to relocate it to a new site out of the city's central business division. The situation is similar in Nakuru and Mombasa and other urban areas.

The Constitution of Kenya 2010 grants the rights to a clean and protected environment along with the associated obligations to protect the environment. The Local Government Act and its successor, the County Government Act—have vested powers to local authorities to establish the necessary systems and procedures that are necessary to deal with SWM at local level (GoK, 2010, 2012). The National Environment Policy 2013, which outlines responsibilities for the government, is well integrated with the National SWM strategy. The Kenya Penal Code makes it an offense to vitiate the environment and its integrity while the Public Health Act focuses on prevention of nuisance that could have negative effect on public health. SWM cuts across various sectors and stakeholders and therefore, policies relevant to SWM are found in a number of sectoral policy documents. The main environment-related policy in Kenya is the National Environment Management Authority (NEMA). The NEMA was established under the Environmental Management and Coordination Act (EMCA) No. 8 of 1999 (EMCA) as the principal instrument of government for the implementation of all policies relating to the environment. The EMCA 1999 and amended in 2015 was enacted against a backdrop of multiple sectoral laws dealing with various components of the environment, with the objective to bring harmony in the management of the country's environment.

As stated earlier, solid waste remains one of the biggest environmental challenges in Kenyan cities as well as their fast growing peri-urban and informal settlements. As with other sub-Saharan countries, access to modern and clean energy is still low in Kenya with 70% of the population still relying on traditional biomass—charcoal and firewood. There are some local community examples of WtE. For example, SimGas enterprises have reached over 800 consumers in Kenya and Tanzania with provision of community biogas digesters. Here, customers collect their own organic waste (e.g., animal dung, kitchen waste, human feces, and agricultural residues) and feed them into the mobile biogas systems provided by the enterprises. Some enterprises provide solutions at community level, where organic wastes are treated to generate biogas for a several households. For example, Skylink Innovators enterprises in Kenya sold over 200 domestic biogas plants and about six large-scale ones, and in the process over 5200 people have benefited. WtE is an option that will not only help to manage solid waste but also improve access to clean energy through WtE conversion technologies to produce electricity and biogas. The process also has potential for other products such as diesel fuel from plastic waste and fertilizer production. These techniques are being implemented at various scales in many countries, but Kenya is yet to exploit this vast resource to its advantage.

Many WtE opportunities are carried out small scale; including production of biomass briquettes for domestic and industrial heating. The briquettes are made from food, agricultural, and forestry-related wastes; such household food wastes, wood residues, and sugarcane bagasse. There is neither national program nor policy guidelines to unlock the potentials of WtE but there is a growing interest in the last few years. A number of feasibility studies are being conducted in the major city, however, none has transformed to mainstream business enterprises.

6.4.4 WtE Project in Uganda

Albert Rugumayo

Faculty of Engineering, Ndejje University of Uganda, Kampala, East Africa

In Uganda, local governments are obliged under the Public Health Act 1964 to provide for collection, haulage, and safe disposal or treatment of solid waste in areas of jurisdiction (NEMA, 2006; Chanakya et al., 2009). The rapid urbanization and lack of investment in appropriate waste management infrastructure and a lack of management and technical skills led to a strategic framework for reform in 1997 and the Kampala City Council (KCC) enacted the SWM Act in 2000 in order to institutionalize private sector participation (Katusiimeh et al., 2012; Nyakaana, 1997; Okot-Okumu, 2015). Currently, solid waste collection in all municipalities of Uganda employ a combination of systems such as the door-to-door collection system; commonly used in high- and medium-income areas, and/or a central collection system mainly used for low-income areas. These systems are implemented either by the private or public operators. The door-to-door waste collection scheme, with collection from households, business premises, and institutions comprises the formal collection. It employs the use of motorized collection to retrieve waste from paying customers. For example, in Kampala, >35 private companies are engaged by the KCC Authority to collect solid waste from private homes or institutions, with the cost for the services being levied to the individual homeowners and institutions (Katusiimeh et al., 2012; Okot-Okumu, 2015; Kinobe et al., 2015). The informal collection system on the other hand involves members of the community and some individuals employed by the formal sectors that pick valuable waste. Although there are no controls, the role of the informal sector in the SWM system cannot be underestimated given their involvement in the recycling and reuse businesses.

In Kampala city, 21% of households are served by informal waste collectors, 44% by the formal private sector, 1% by CBOs, and the rest are managed by KCCA (Okot-Okumu, 2015). Within these systems, municipal authorities independently or through private-public partnerships also undertake collection of solid waste mainly from markets, public places (business centers), community transfer points (skips, bunkers, and open-road verges), some households, or at a few agreed locations (Okot-Okumu, 2015). Records at Kiteezi dumping site in 2009 indicated that KCCA alone was responsible for 87,000 (about 27%) tonnes of garbage dumped compared to the private sector's 237,000 tonnes (73%) (Katusiimeh et al., 2012). However, in spite of the shift toward the involvement both public and private sectors in SWM, there are still major problems in the delivery of waste management services (NEMA, 2006). For instance, although SWM costs to municipalities and towns constitute between 20% and 50% of municipal revenues, between 50% and 70% of the residents benefit from waste collection service (Nyakaana, 1997). Studies indicate that in Kampala, about 35%–40% of the 1500–2000 tonnes of solid waste generated per day is collected and transported to the official dumpsite at Kiteezi (Mugagga, 2006; WaterAid Uganda, 2011; Komakech et al., 2014). This implies that 60% of the garbage generated daily is not collected and disposed of. Uncollected waste is either indiscriminately disposed (58.7%), dumped in designated disposal areas (15.5%), disposed of in backyards (10.1%) or buried and burned, used as feed to livestock and home composting, etc. (15.7%) (Okot-Okumu, 2015). For instance, in a study carried out in the Busia Municipal Council in 2015, it was noted that only 50% of the 10,300 tonnes of MSW generated were disposed of by means of formal collection, whereas in some areas, limited access to waste collection enhanced littering, backyard dumping, and burning of wastes (Lederer et al., 2015).

It is worth noting that the efficiency of the collection system is tagged to waste being delivered to a central designated/community transfer points. Skips (with a capacity of 5 to up to 20 m^3) and bunkers (three sides, with an open roof) often located along the main roads, next to markets and in most commercial areas are the most common community collection/transfer facilities. Unfortunately, these are inadequate to meet the waste generation rates in municipal areas. This coupled with irregular collection schedules results into leaching of rotting waste into the environment, scattering of waste by animals, wind (e.g., paper and plastic bags) and scavengers, or waste clogging storm drains and channel causing flooding. In addition, the highest collection levels in Uganda's municipalities are realized in a few vital areas such as marketplaces, upscale residential areas, and politically sensitive areas served by private companies (formal collection) (Tumuhairwe et al., 2009). The densely populated urban areas with poor accessibility, inhabited with the urban poor communities who are unable to pay for collection services, receive the lowest waste-collection services. In these areas, the informal waste collectors (often members of the community) pick waste from homes (usually in waste bags), community transfer points (skips and bunkers), and unofficial dumps (e.g., road verges), thus accelerating the problem of indiscriminate dumping of waste, since they are not under any form of control/monitoring.

The transportation of waste maybe carried out formally by the municipal or town councils, and by private companies licensed and under contract to the municipalities to deliver wastes to gazette disposal or treatment points. Informal transportation modes are individually organized mainly to deliver waste from the generation points to community transfer points, pits, or other places, such as illegal sites. Waste-transport vehicles common to both the public (municipalities) and private (licensed companies) include; covered compressor trucks, open trucks, and trailers, skip carriers. Common issues with transportation include; failure to provide personal protection gear for workers and to cover the trucks to avoid littering, odor, and aesthetic problems (Chengula et al., 2015; Okot-Okumu, 2015; Henry et al., 2006).

When solid waste is collected, it is transported to locations where it is, in one way or another, treated and/or disposed of. In Uganda, solid waste is formally disposed of in gazette waste-disposal point or by land filling. Waste may also undergo treatment through composting and energy recovery. In Kampala city, private companies hired by the KCC Authority for management of waste disposal are paid a monthly fee, based on the tonnage for garbage delivered and disposed of, whereas the private collectors pay a subsidized disposal fee of UShs 2000 per ton disposed at the landfill to the companies (Office of Auditor General, 2010). An increase in garbage collection in Kampala was realized after additional budget allocations were made. It is estimated that $10,654,811 per year is spent on solid waste collection in Kampala alone (Nabukeera et al., 2014). However, other municipalities without adequate resources, the amount of solid waste generated daily overwhelm their collection capacity.

According to the Uganda State of the Environment Report (NEMA, 2006, 2014), 6.4 million people live in urban centers in Uganda, and with increasing urbanization, this numbers are expected to increase to over 30 million by 2040. This trend, coupled with increased productivity, resource consumption, and economic development, will result in increased levels of solid waste generation. It has been estimated that <40% of the solid waste generated in the urban cities of Uganda is collected for recovery or disposal. The remainder is illegally dumped in drains, water bodies, or in the open dumps (Komakech et al., 2014).

The waste management system is poorly coordinated and refuses collection vehicles spend several hours traveling to and from the dump sites each day due to traffic congestion in the cities. This increases vehicle collection time, which results in inconsistent collection schedules with uncollected waste and litter on the streets. This leads to increased flooding during the rainy season and increased risk of vector-related diseases and impacts to water resources. The solid wastes generated from Uganda's municipalities emanate from diverse sources that include amongst others; households (residential), commercial premises, markets, institutions, industries, and health-care facilities. All urban centers are also experiencing unprecedented population growth rates since 2002, the data on SWM at all levels from the household, up to the municipal level is scanty, or if available at all, it is generally unreliable, scattered, and poorly organized. The total amount of solid waste generated by urban centers varies according to the location and season; it has been estimated that the average solid waste generation rates are 0.55 kg/capita/day; low-income municipalities generate about 0.3 /kg/capita/day, compared to the higher-income municipality rate of 0.66 kg/capita/day (Ojok et al., 2013; Lwasa et al., 2007). Lira and Kampala are reported to have generate rates of 0.3 and 0.5 kg/capita/day, respectively, while the City of Hoima generates 0.2 kg/of waste per capita/day [Urban Research and Training Consultancy E.A Limited (URTC), 2012]. For instance, Kampala city is currently the most populated urban center in Uganda with a populace of approximately 1.5 million people, and this generates approximately 30,000 tonnes of waste per month.

In 2000, the solid wastes generated in the urban areas of Uganda was reported to typically consist of 73% organic matter, 5.4% paper, 1.7% sawdust, 1.6% plastics, 3.1% metals, 0.9% glass, 8.0% tree cuttings, 5.5% street debris, and others constitute 0.8% as shown in Table 6.3 (NEMA, 2006). The report further indicated that the vegetable/organic matter content in wastes would reduce to 30% over the next 20 years with substantial increases in the composition of metals, plastics, and glass.

Table 6.3 The Composition of Solid Waste Generated in Urban Areas by Percentage, Including the Projected Percentages to 2020

Material	Year 2000	Year 2010	Year 2020
Vegetable/organic matter	73	50	30
Paper	5.4	21.1	30.9
Sawdust	1.7	1	1
Plastics	1.6	6.3	8.8
Metals	3.1	12.1	17.9
Glass	0.9	3.5	5.7
Tree cuttings	8.0	4.0	4.0
Street debris	5.5	2.0	2.0
Others	0.8	–	–

Data from NEMA, 2006. State of the Environment Report for Uganda 2006/2007. National Environment Management Authority, Kampala, Uganda.

Table 6.4 Composition of Solid Waste Generated by Municipalities in Uganda

Composition	Fort Portal	Jinja	Kabale	Kasese	Lira	Mbale	Mbarara	Mukono	Soroti	Kampala
Food	36.6	31.9	40.5	49.8	36.4	31.9	55.6	28.8	28.3	58.2
Garden	36.1	36.7	29.6	24.2	32.3	36.0	24.5	46.2	37.7	22.1
Paper	6.8	8.0	5.2	5.4	5.5	7.5	2.6	5.7	7.2	6.1
Plastic	8.4	7.9	8.1	5.1	6.8	10.8	4.7	7.9	8.8	7.2
Glass	0.7	0.7	0.5	0.4	1.9	0.9	0.6	0.4	0.7	0.7
Metals	0.0	0.5	0.5	0.1	2.2	1.0	0.2	0.3	0.4	0.3
Textile	1.0	1.8	1.8	0.5	1.2	1.0	0.3	0.4	2.5	1.8
Soil, ash, stones and debris	10.2	12.5	13.7	14.7	13.7	10.8	11.5	10.2	14.4	3.6
Others	9.8	9.0	10.1	13.4	12.2	7.7	9.9	8.4	11.1	2.6

Data from NEMA, 2006. State of the Environment Report for Uganda 2006/2007. National Environment Management Authority, Kampala, Uganda; Mohee, R., Simelane, T., 2015. Future Directions of Municipal Solid Waste Management in Africa. Africa Institute of South Africa, Pretoria, South Africa.

A 3-year study revealed no significant change in the organic/biodegradable component of the waste generated over the years in selected urban centers of Uganda (Okot-Okumu, 2015). Waste generated in the municipalities still has a high organic content (73%–83%) as shown in Table 6.4.

Studies carried out by Komakech et al. (2014) to characterize Kampala City's MSW at the Kiteezi landfill revealed the following composition (by weight): organic material (92.1%) with smaller percentages of hard plastic (1.8%), metals (0.1%), papers (1.3%), soft plastic (3.0%), glass (0.6%), textile (0.5%), and 0.6% leather (Nabukeera et al., 2014; NEMA, 2014; KCCA, 2017). The variances noted in the organic composition of waste delivered during the dry compared to wet has been attributed to the abundance of food crops in the city's hinterland and peri-urban areas that occurs in the harvest season, which coincides with the dry months. On average, therefore, of the waste generated in KCC has a 70%–80% organic waste and a moisture content of 71% (KCC, 2006; Kinobe et al., 2015)—shown in Fig. 6.10.

In view of the above composition of the urban council wastes, the solid waste generated can generate energy in Uganda through thermochemical, physicochemical, and biochemical processing. A major challenge with thermochemical processing is the high moisture content that would greatly hinder the combustion process raising the ignition temperatures (Kathirvale et al., 2004) and the considerable amount of inorganic waste therein that would need to be removed (Byrne et al., 2015). Nonetheless, other reuse options such as composting at both a small and commercial scale need to be promoted considering that Uganda is an agricultural country. In addition, to reduce on the amount of waste that is land-filled, the other nonbiodegradable materials of economic value such as metal and glass ought to be gathered, sorted, and reused or recycled (Ekere et al., 2009; Banga, 2011).

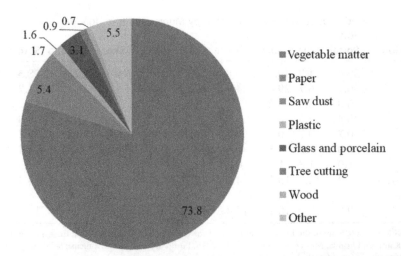

Fig. 6.10 Composition of solid waste generated in Kampala City Council. *(Data from Nabukeera, M., Boerhannoeddin, A., Ariffin, R.N.B.R., 2014. Division solid waste generation and composition in Kampala Capital City Authority, Uganda: trends and management. IOSR J. Environ. Sci. Toxicol. Food Technol. 8(10), 57–62; NEMA, 2014. National State of the Environment Report for Uganda 2014. Harnessing our Environment as Infrastructure for Sustainable Livelihood & Development. National Environment Management Authority, Kampala, Uganda.)*

The volume of solid waste generated in urban centers in Uganda has increased over the years from 87.7 to 645.5 tonnes of waste per day in 2006, to >1500 tonnes/per day in 2016. For instance, in Kampala city, waste generation increased from 198 tonnes/day in 1969 to 1200 tonnes/day in 2004 (Okot-Okumu, 2015). This trend has been mainly attributed to rapid urbanization, industrialization, and population growth; the disproportionate and limited capacity of the institutions in the management of solid waste led to increased indiscriminate disposal of waste into the environment (Mohee and Simelane, 2015; Banga, 2011). Presently, waste generation in Uganda ranges between 1.2 and 3.8 kg/day (Kinobe et al., 2015). According to Okot-Okumu and Nyenje (2011), greater quantities of solid waste are generated from residential areas or households and markets than from commercial areas. Conversely, densely populated urban zones (e.g., slums) characterized by low-income households, have low waste generation rates ranging between 0.22 and 0.3 kg/cap/day compared to higher-income households that generate 0.66–0.9 kg/cap/day on average (Kinobe et al., 2015). Generated wastes are delivered to transfer/collection points, and then they are collected by urban council workers or private operators; finally, transported to designated locations for treatment or disposal. In Uganda, waste is formally disposed of at a waste-disposal point or landfill. Waste may also undergo treatment through composting and energy recovery. According to KCCA (2017), the quantity of solid waste delivered to Kiteezi landfill is approximately 1300 tonnes per day, of which 90% is landfilled (spread, compacted, and covered with soil) while 10% is collected by scavengers. Reports indicate that due to unrestricted access to the landfill, >500 scavengers flock the dumping site daily in search of recyclables materials, ranging from plastics, scraps, and clothes to food leftovers. The valuable recyclables are often sold off to private intermediaries and plastic recycling companies. Although the quantity of generated waste that is reused/recycled in Kampala is not known, it is likely to be close to the 11% value, which is cited for Dar es Salaam city (Komakech et al., 2014). Following the decomposition process, approximately 166 m³/day (61,000 m³/year) of leachate that flows from the landfill, is collected via high-density polyethylene (HDPE) 160-mm diameter pipes, and directed to the leachate treatment plant where it undergoes treatment by aeration (equalization, settling, aeration, and clarifying processes) (Okot-Okumu, 2012; KCCA, 2017).

Outside Kampala, there are several Clean Development Mechanism (CDM) projects funded by the World Bank Partnership that promotes composting of organic wastes. The project started in nine municipalities (Lira, Soroti, Mbale, Mukono, Jinja, Fort Portal, Kasese, Mbarara, and Kabale) in 2005 and was further expanded to eight more municipalities (Masindi, Busia, Hoima, Mityana, Entebbe, Arua, Gulu, and Tororo) in 2012. The composting projects' overall aim is to reduce methane emissions from MSW especially while at the same time producing manure for agriculture (NEMA, 2014). Solid waste is delivered by truck to the compost plants, and although there no weighbridges to establish the total waste handled at the facilities, the projects are assumed to be working at an optimal design capacity of 70 tonnes/day. In addition, energy recovery from waste is ongoing within the country at varying scales but not on the municipal scale. For instance, *bagasse* a waste by-product in sugar factories is burnt at a high temperature to produce steam, which in turn drives turbines to generate electricity for local consumption in the sugar industry, and domestic and industrial wastes are used to make biomass briquettes. The making of briquettes for use in energy stoves using biodegradable organic waste such as banana peelings, coffee and rice husks, etc. is being implemented by different interest groups across the country (Solomon, 2011; Okot-Okumu, 2015). However, despite these innovations, there is generally a lack of data on the quantities of solid waste being used for these operations.

WtE enterprises in Uganda are typically small scale, domestic installations with bioenergy products like biogas and carbonized briquettes for cooking fuels (Menya et al., 2013) since an estimated 72.7% of the population use traditional cooking stoves (Okello et al., 2013). The current use of inefficient cooking stoves is also partly responsible for indoor air pollution and respiratory illness reported amongst its users (Tumwesige et al., 2015). Modern bioenergy technologies such as biomass gasification, cogeneration, biogas generation, biomass densification, and energy-efficient cooking stoves are opportunities to improve energy supply systems, but have not been widely disseminated in Uganda (Okello et al., 2013). The household biogas plants in Uganda typically use agricultural waste, for example, cow dung and the biogas plants built at institutions (e.g., schools and health centers) have used a combination of agricultural wastes and human waste (with connection to a toilet/latrine). Usually, for most of the biogas plants in Uganda, the gas generated is used for cooking and to some smaller extent for lighting purposes. By 2009, over 600 household digester plants had been installed in Uganda over 20 years since its inception in the country (Tumwesige et al., 2014; Walekhwa et al., 2010). The widespread adoption of biogas systems has been hampered by lack of adequate information on its production, economic viability and potential benefits and prohibitively high capital costs, and lack of water for most households in the rural areas that do not have piped water. The key problems reported are: poor maintenance, poor construction or design leading to gas pressure problems, high maintenance costs, and weak or no technical support (Bond and Templeton, 2011). However, studies from biogas plants in Eastern and Central Uganda revealed that investing in household biogas digester plants is economically feasible in terms of payback period, net present value, and internal rate of return (Walekhwa et al., 2010). Generation of electricity from biogas is still limited in Uganda today, in spite of the availability of the abundant feedstock for anaerobic digestion—such as MSW, livestock waste, and municipal/industrial wastewater. However, a study by Oyoo et al. (2014) has shown that integrating anaerobic digestion, resource recovery, landfill of MSW, and sewerage in Kampala (the capital of Uganda) can reduce many of the environmental impacts from wastes. The assessment carried out in Jinja, the second largest city in Uganda, revealed that the available OFMSW resource could provide about 10% of the total electricity required in the city (Mark, 2012).

Several biogas-to-electricity projects in Uganda are underway. For example, a slaughterhouse in Kampala treats its waste and wastewater through anaerobic digestion to produce methane (biogas), which is converted to electricity that is used to power 15 security lights, 15 deep freezers, and 15 refrigerators at the abattoir; and saves an estimated US$2800 per month through avoided electricity purchase (Mugalu, 2015). A new sewerage plant constructed in Bugoloobi, Kampala is expected to produce biogas that will generate about 650 kW of electricity that is sufficient to power the plant (Otage, 2016). In Gulu District (northern Uganda), it is planned to generate biogas from agricultural wastes (maize silage and cow dung) with a total capacity of 1 MW electricity and 600 kW thermal energy; as well as producing approximately 4000 ton of organic fertilizer/year. The generated electricity will serve a primary school and a health clinic, three industries (a briquette plant, a drip irrigation company, and a cold storage/milk chiller/processing plant), and >3500 people (Krishan, 2012).

Biomass densification is a quite a common practice in Uganda, where there is rampant use of wood fuels (firewood and charcoal) to provide energy for cooking, and is typically directly burnt in open fires or stoves. Over 80% of households in urban areas depend on solid fuels such as firewood and charcoal, which are burnt in stoves (Ferguson, 2012). Biomass densification typically involves the use of biomass crop residues and other organic wastes for producing biomass briquettes that are used instead of charcoal or firewood. Switching to briquettes from waste is easy since this does not involve changing the burning system and about 20% of households are already using densified wastes, either in carbonized or non-carbonized form. Production of fuel briquettes involves the collection and compaction of a combination of combustible wastes and processing them into a solid fuel product of any convenient shape that can be burned like wood or charcoal. For example, in Namatala slum (Mbale district in eastern Uganda), doughnut-shaped briquettes of 4 in in diameter are made from discarded coffee hulls, rice husks, charcoal particles, sawdust, wood chips, and waste paper. The produced briquettes are sold to different users (Njenga et al., 2009). Briquette manufacturers can be categorized into: microscale (<2 tonnes/year)—who produce briquettes manually such as at a household level; small scale (<200 tonnes/year)—use manual-aided equipment in production; medium scale (<2000 tonnes/year)—use automated machines such as Kampala Jellitone Suppliers (KJS); and large scale (<20,000 tonnes/year)—use automated machines to manage large amounts of feedstock. Large-scale producers are not available in Uganda. However, the Uganda Investment Authority put forward investment proposals for a 70 tonnes per day manufacturing plant for briquettes made from MSW collected from households and surrounding markets in Kampala, where about 1500 tonnes of MSW are produced daily (Ferguson, 2012). Increasingly, medium and large-scale producers are making use of other biomass residues such as agricultural wastes and MSW to produce briquettes. Several businesses use the abovementioned feedstock, but the MSW resource remains largely untapped in Uganda (Mawatha and Lanka, 2015). The recent rise in the number of briquetting businesses is a good indicator of its market feasibility in Ugandan markets. This is mainly driven by increased competition due to recent price hikes of the commonly used fuels like charcoal; and this is reflective of increasing consumer interest and awareness of briquettes as a competitive source of fuel. Briquettes can be retailed at super markets in town areas and other institutional and industrial users, but this is hampered by the lack of a wide retailer network (Mawatha and Lanka, 2015). KJS is Uganda's first and largest producer of carbonized and non-carbonized briquettes. KJS annually produces about 2000 tonnes of briquettes made from agricultural wastes; such as sawdust, peanut husks, and coffee waste for cooking and heating in households schools, universities, hospitals, and factories.

In Uganda, incineration is mainly limited to use of small-scale incinerator at institutions such as hospital and hostels, without energy recovery (Miyingo, 2012). There is a significantly greater potential, as demonstrated by the city of Addis Ababa, Ethiopia, where the first incineration plant in Africa has been established—it will burn over 1400 tonnes of waste per day and generate 55 MW electricity. While there is currently no large-scale incineration of waste for energy recovery, in Uganda, the government has received the support from the Global Enforcement Facility and is set to construct five WtE plants in the districts of Arua, Jinja, Mbale, Mbarara, and Masaka; that will generate 2.9 MW electricity (Othiang, 2018).

CHAPTER SEVEN

Existing Barriers for WtE in Developing Countries and Policy Recommendations

Suani Teixeira Coelho
Research Group on Bioenergy, Institute of Energy and Environment, University of São Paulo, São Paulo, Brazil

7.1 BARRIERS FOR WtE IN LATIN AMERICA

Walter Ospina, Enrique Posada Restrepo, Atilio Armando Savino, and Suani Teixeira Coelho

Consejo de Investigación y Tecnologías de Valorización Energética de Residuos de Colombia (WTERT—Colombia), Colombia, SC, United States of America
Área de Innovación y Desarrollo Hatch, Portland, OR, United States of America
International Solid Waste Association (ISWA), Buenos Aires, Argentina
Research Group on Bioenergy, Institute of Energy and Environment, University of São Paulo, São Paulo, Brazil

The introduction and implementation of waste to energy technologies will be possible if the different barriers that exist could be solved. The first thing to do is to identify such barriers and propose some actions for their management and mitigation, with the purpose of promoting the realization of sustainable and well-designed projects of WtE systems that become truly beneficial for the region.

7.1.1 Technical Barriers in LA Countries

7.1.1.1 Lack of Knowledge of WtE Technologies

The current mind-set in the region is to take the wastes to final disposition in open dumps or controlled landfill sites. There is not enough public awareness of waste to energy (WtE) systems as a desirable treatment option. There is, in general, lack of knowledge, on the part of public officials, the municipal and regional authorities, and entities responsible for the management of MSW. Of course, people know in general the subject matter, but they do not consider WtE solutions as a practical, readily applicable ones, as compared to improve and enlarge landfill systems, impulse recycling options and propose relatively timid schemes to separate on the origin and developing some composting and separating mixed waste. The current emphasis is having better transporting systems from the sources to the landfill sites, including intermediate transfer stations.

Still, there is starting to appear in different large metropolitan areas the idea of applying WtE solutions, which indicates that these possibilities are becoming to be considered by authorities and planners. However, there is no solid information program to communicate to the different stakeholders the benefits and possibilities of applying these kinds of technologies.

Municipal Solid Waste Energy Conversion in Developing Countries
https://doi.org/10.1016/B978-0-12-813419-1.00007-3
219

7.1.1.2 Lack of an Appropriate Technological Base and Lack of Complete Engineering

In general, most countries in LA region do not have a portfolio of companies that can manufacture equipment for thermal treatment systems capable of handling hundreds or thousand tons per day of mixed waste, burning them in a controlled way, generating electricity, and controlling the air pollution problems related to this. So, there are no local suppliers pushing for projects or offering alternatives as Engineering Procurement Construction (EPC) installations. This means that local responsible waste-handling entities will tend to look for solutions with external providers unless some kind of industrial and manufacturing development could be put into place locally. It is important that concerned authorities and planners see the opportunities for job and technology creation if a strong commitment is assumed in WtE solutions. An important example is what has happened in China and India, which have developed competitive sectors in WtE technology, able to confront their own situations and to export technology and equipment.[1]

Engineering and design are very important components of the technology necessary to impulse WtE in a country. These systems require detailed studies and planning activities and it is advisable to do the projects taking into account all the engineering stages. There is always the temptation and the idea that the projects can be accelerated and put into place based on the experience and support of suppliers and makers, by means of EPC developments, in such a way that engineering stages can be simplified or even avoided. This normally is a much costlier and rigid solution and does not contribute to developing local technology and prosperity. In the solution of the problems, there is ample space to develop a region, as compared to relying only on external provided solutions.

Arriving at a hypothetical implementation case, the lack of experience of local human resources for the operation of the plant will add an extra challenge.

Another experience is in Brazil corresponds to the two WtE plants under development, as mentioned above: the 20 MW (865 t/d of MSW) incineration plant under construction by Fox Co. with different partners in a PPP; the small scale 1 MW MSW-gasification plant in Boa Esperança municipality (55 t/d of MSW). From these two experiences, other small municipalities are quite interested in the WtE implementation, contributing to the dump shutdown.

7.1.2 Political Barriers in LA Countries

7.1.2.1 Interests and Influence of Existing Waste Managing Concessionaires

The companies that collect, transport, and operate the dump or sanitary landfill in municipalities, being large, medium, and small, will feel themselves negatively affected or

---

[1] For more information on these topics, please visit:https://www.bioenergyconsult.com/swm-outlook-india/; https://www.bioenergyconsult.com/waste-to-energy-china/; https://www.newsecuritybeat.org/2017/11/default-post/; https://economictimes.indiatimes.com/industry/energy/power/how-to-make-waste-to-wealth-a-reality/articleshow/62490388.cms.

menaced when somebody proposes a WtE project to be integrated into the SWM system. This, taking into consideration that this new system could mean less as usual business and even the possibility that the concession contract disappears. In this sense, municipalities that do not have concessions to collect, transport, dispose, and operate the dump or sanitary landfills have more freedom to undertake a WtE project.

These conflicts of interests and negative consequences must be specially considered as they are legitimate. There must be possibilities for the existing concessionaries to become part of the new WtE schemes, especially as investors, considering the large capital requirements for these projects. Another consideration is to introduce the WtE projects as a growing part of the total solution, considering that it would be quite difficult to change the existing systems at once. Also, the WtE systems will generate new opportunities and new areas of work for concessionaires in aspects related to collecting, separating, classifying, and valorizing waste.

In Brazil, there are now only the two WtE systems mentioned above: the MSW-incineration process in Barueri, Sao Paulo State, and the one MSW in Boa Esperança, Minas Gerais.[2] The incineration plant is developed by Fox Co.[3] in a PPP with the partnership with Keppel Seghers.[4]

7.1.2.2 Barriers Related to Political-Technological Relationships

Putting into practice a WtE system will involucrate a series of interrelations between political and technological interest and groups, such as politicians, elected legislators (Congress, Assemblies, Municipal Councils), executive functionaries at the national, regional, and municipal levels, and functionaries of the environmental and sanitary authorities. All of these relate to the technical sector represented by the engineering guilds or by technical consultants and technical groups of interest.

Before the appearance of the Internet, the powerful search engines, and digital networking commercial and technological systems, politicians and public officials in charge of originating public projects, used to keep closer communications with the professional guilds of engineers and architects; this would help defining priorities in projects and technologies for the municipality.

It should be pointed out that engineers and architects are, by nature, promoters of projects and that politicians and public leaders execute projects as a natural result of their management. Therefore, their common relationships are beneficial for the society. These relationships seem today more distant. They tend to appear only when one of the parties has a specific need. Professional engineering societies look to politicians when they require that some kind of regulation be created and approved. In the opposite direction,

---

[2] In Boa Esperança, the MSW gasification process was developed by a local manufacturer (Carbogas Co., www.carbogas.com.br).The Boa Esperança plant is under development in an agreement of Furnas Co. and Carbogas.

[3] http://haztec.com.br/solucoes-ambientais-completas/index.php/solucoes/unidades-de-recuperacao-energetica.

[4] http://www.keppelseghers.com/en/content.aspx?sid=3033.

politicians could look for organized engineer society's support when some problems arise in the execution of projects, for example, failures, accusations of mishandling, cost over-runs, and serious breaches in the deadlines, to help explaining and clarifying the situations. In a few cases, the guilds of engineers and architects are officially consulted.

For the developing of more sophisticated waste management systems, as is the case for WtE, it is convenient to establish strong connections between engineering (represented by professional societies and guilds) and politicians and policy makers. In the final run, it is important that engineering has the respect and recognition of politicians. For this, it is necessary for the professional societies and groups to reach maturity and to develop capabilities to study and propose projects and alternatives and to exercise good communications to be able to listen and convince.

7.1.2.3 Barriers Related to the Legal and Institutional Framework for WtE Projects

There is a general framework that opens the way for WtE to be considered as systems that employ renewable energies. However, there is still a lack of a specific legal framework to generate renewable energy from waste. As it can be seen, the regulations are to be generated by different ministries and entities in the national and regional government, which seem to be slow and lacking enough interest to act speedily. This is an obstacle for institutional investors, such as municipalities and governorates, to execute WtE projects. It would be appropriate to simplify the regulatory process following the already existing developments on WtE systems in several countries.

It is necessary to analyze if there are any restrictions in the custom regulations that avoid the import of different parts used in a certain technology.

One of the obstacles to overcome is the developing of clear terms of reference for calling for a necessary bid to select the provider.

A weak institutional framework is a common characteristic of the different countries of the region. A good example of that is one of the most difficult barriers to defeat: the change of attitude of a new government. It is very common that the new government will change or paralyze the previous decision.

An interesting experience of an adequate framework for other countries comes from Portugal. Valorsul Co.,[5] operating with MSW from Lisbon and five other municipalities, produces 37 MW from 2000 t/day. The huge investment (9 million €) was possible due to two factors[6]:

- the special "green" tariff for the electricity generated (84 euros/MWh).
- the donation of 54% of the total investment by the European Union.

---

[5] http://www.valorsul.pt/en/seccao/areas-de-negocio/energy-recovery/.
[6] Personal communication during technical visit (S. Coelho, several years).

7.1.2.4 Barriers Related to Zero Waste Policies

The programs that some countries, organization, municipalities, and departments (provinces) have established under the slogan "ZERO WASTE," following international leads, must be accompanied by tax and coercive measures, so that recycling increases and discourages the use of dumps and landfills. Since the environmental authorities do not regularly apply such measures, the municipalities prefer to continue implementing landfills or sanitary landfills, considering it cheaper than any other alternative, letting future generations solve the resulting social and environmental problems at a higher investment cost. In short, our countries tend to promote written regulation, at times complex, at times well intended, but without either tools or real intentions or procedures to apply such regulation. This constitutes a barrier not only for WtE projects but also for any type of projects. It happens that public officials and organizations that promote zero waste programs do not include WtE projects, as part of them, as it happens in Europe, Japan, China, and Korea. There must be knowledge and practical wisdom to understand the ways of reaching minimum waste generation and minimum waste disposed improperly into the environment, as absolute Zero waste is impossible. WtE is perhaps the best practical advance in the right direction.

7.1.2.5 Barriers Related to the Availability of Land

In large metropolitan areas, the population density implies that it is very difficult to find a place to install a WtE plant. Also, it is common to find very strict regulations to locate a waste management facility in a very dense city in case the normal traffic circulation is affected.

It is worth to mention that very modern WtE plants are placed in the very center of very well-known developed cities like Paris, Vienna, Hamburg, Copenhagen, etc.

7.1.3 Economic Barriers in LA Countries

7.1.3.1 Barriers Related to the Required Initial Investments

When public leaders and officials, at the national, regional, or municipal level understand that there are social, economic, and environmental benefits associated with WtE projects, they see as an imposing barrier the high initial investment required to develop these projects. Usually, this is argued without complete knowledge and without a good-quality prefeasibility study previously prepared to really know which is the most applicable technology and the investment cost of it, considering the totality of the costs and benefits. To pretend that any project will recover its initial investment, considering only the sale of by-products, such as electricity and bio-fertilizer, in a few years, is not feasible. This consideration does not include in the evaluation other social, economic, and environmental benefits of the project when it avoids X ton/day of waste to be deposited in an open dump with all the negative impacts and costs.

Fortunately, manufacturers of equipment for WtE plants are making technological advances that lower the prices of WtE technologies, allowing them to be paid by developing countries when correct cost-benefit analysis is done. Another consideration is the convenience of establishing local technological, manufacturing, and engineering capacities to be able to execute these projects locally, to create jobs and prosperity, and diminish investments for each project, eventually being able to export technology and equipment to other countries.

It is important to develop procedures to demonstrate from the technical, environmental, and social points of view, the advantages and real benefits, and costs and initial investments for a WtE project. This must include a critical analysis of the real costs, associated investments, and life-cycle environmental impacts of sanitary landfills. This could help to better define the MSW projects required by a municipality, region, and the country.

7.1.3.2 Barriers Related to the Interests of the Electricity Sector and Their Perspective of WtE

The waste sector must consider a broader perspective and see electricity generation as one aspect of WtE and not the only goal, considering the benefits from the relationship of WtE systems and basic sanitation. The interests of the electricity sector and the waste management sector, apparently, are not the same and do seem to coincide. But there is a common ground to consider this is sustainability and the need to integrate solutions looking for the real (evident and hidden) cost and benefits to society.

Here also appears that the matrix of electricity costs is different in each country and then the final economic equation to close any project will be affected.

In general, the price of energy generated by WtE technology is not competitive with the other sources of renewable energies. It is common to analyze the possibility to guarantee a fixed subsidized price for the project time of implementation.

7.1.3.3 Barriers Related to Tipping or Gate Fees

These are the fees that landfills and treating plants (such as WtE plants), charge for each ton of MSW they receive to make the final disposal. The tipping fee of a WtE plant should complement the sale of electricity to repay the investment and operational costs.

For the economic feasibility of this kind of project, it is very important to know how much money you save for the change of the final disposition.

For that reason, it is very important to know the exact cost of each ton of waste that we diverted from a sanitary landfill to a WtE plant.

Many times the tipping fee of the sanitary landfill is determined without considering the total cost for political reasons or incorrectness in how they are calculated.

Sometimes the landfill has been built in a municipal land and then the amortization of the valuation of it is not considered. In other cases, the funds that should be created to buy another land when the landfill is completed are also not considered.

Finally, it is not always considered the costs of the operational closure and maintenance of the landfill.

The case is worse because in many countries it is considered that the municipalities should not pay this service.

7.1.4 Public Perception About WtE Technologies in LA Countries

7.1.4.1 Barriers Related to the Not in My Backyard (NIMBY) Effect

The effect is very well known as the rejection of part of the community involved in the vicinity of a waste management facility.

The arguments used are the disturbance caused by the increased transport, the disturbances created during the day, especially at night, odors, and the most important one is the perception that the health will be at risk.

7.1.4.2 Experiences Related to Public Perception

In Brazil, for instance, there is a significant concern from local society, related to WtE technology, mainly due to the fear of pollutant emissions (dioxins and furans) from incineration systems. There is not enough information about strict standards defined by the local environmental legislation, as well as about the existing gas cleaning systems to achieve such standards. Interesting to note that this behavior is similar to those in other countries, such as in Portugal, when WtE Valorsul has faced several criticism from the local population; and it was realized that, after the construction of the Valorsul plant, there was no more criticism was noted.[7] This indicates the need for adequate dissemination of information to the local society.

On another hand, the same information dissemination is necessary in the case of MSW gasification, since people are not aware that the gasification process does not form dioxins and furans due to the absence of oxygen.[8]

In addition, in Brazil, there is a public concern related to the possible impact of WtE technologies on existing jobs. In most Brazilian municipalities there are a significant number of unskilled workers (called "*catadores*") in charge of the waste separation for recycling. Therefore, there is the fear that WtE technologies will not allow anymore this work and all these workers will lose their jobs. Here again, some capacity building is needed to explain that recycling is absolutely necessary before any WtE technology; so, "*catadores*" will be necessary for those municipalities where there is no recycling system introduced (and this is a reality in almost 100% of the Brazilian municipalities).

---

[7] Personal communication informed during technical visits (S. Coelho, several years).

[8] Gasification process is the incomplete combustion and there is not enough oxygen to allow the formation of dioxins and funans.

7.1.5 Barriers to WtE Implementation and Proposed Actions to Overcome Them

Table 7.1 presents a resume of the aforementioned barriers and some actions that may be considered as a way to overcome these barriers. The implementation of these barriers will depend on the boundary or framework conditions in each country and the relevance of each barrier.

Table 7.1 Barriers to WtE Implementation in LAC and Proposed Actions to Overcome Them

Barrier	Proposed Actions
Lack of knowledge of WtE technologies	Encourage knowledge and public awareness among people, academy, designers, consulting firms, authorities, public officials, companies and entities responsible of managing waste
Lack of an appropriate technological base and lack of complete engineering when doing projects	Stimulate local technology and engineering in the projects to develop technology, create jobs and prosperity Promote application and knowledge of all the engineering stages (conceptual, basic, detailed, execution) to each project • Planning and design based on the establishment of clear objectives • Execution under technical criteria • Control and monitoring of execution to be within the budgeted costs and with the required quality • Exercise feedback and recurrent work, based on discipline, interdisciplinary group work, motivation and leadership, to achieve constant perfection
The interests and influence of existing waste managing concessionaires	Stimulate the existing concessionaries to become part of the new WtE schemes, especially as investors, considering the large capital requirements for these projects Introduce the WtE projects as a growing part of the total solution, as it would be quite difficult to change the existing systems at once Allow WtE systems to generate new opportunities and new areas of work for concessionaries
Political-technological relationships	Stimulate strong connections between engineering (represented by professional societies and guilds) and politicians and policy makers

Table 7.1 Barriers to WtE Implementation in LAC and Proposed Actions to Overcome Them—cont'd

Barrier	Proposed Actions
	Professional societies and groups should reach maturity and develop capabilities to study and propose WtE projects and alternatives, to exercise good communications to be able to be listened to and convince
Required initial investments	Develop procedures to demonstrate from the technical, environmental and social points of view, the advantages and real operating benefits and costs and initial investments for a WtE project
	Include critical analysis of the real costs, associated investments and life-cycle environmental impacts of sanitary landfills. This could help to better defining the MSWH projects
Legal framework for WTE projects	Stimulate the completing of the regulatory norms for the existing laws
	Try to stimulate simplifying the regulatory process following the already existing developments on WtE in several countries
Interests and perspectives of WtE as seen from the electricity sector	The waste sector must consider a broader perspective and see electricity generation as one aspect of WtE and not the only goal.
	It is necessary to understand the WtE projects from the economic point of view and add to the economic considerations the corresponding social and environmental evident and hidden benefits, which should be monetized and included in the cash flow to evaluate a given project
Zero waste policies	There must be knowledge and practical wisdom, to understand the ways of reaching minimum waste generation and minimum waste disposed improperly into the environment, as absolute Zero waste is impossible. WtE is perhaps the best practical advance in the right direction
Tipping or gate fees	Some revisions of this should be proposed to help financing a WtE project, understanding that gate fees complement the sale of energy a way of finance this type of project, which require high investments and requires public equity

Continued

Table 7.1 Barriers to WtE Implementation in LAC and Proposed Actions to Overcome Them—cont'd

Barrier	Proposed Actions
Availability of land	Make use of the advantage that WtE projects have in relation to land use, as compared to sanitary landfills, through public audiences
NIMBY effect	The stakeholders that should be considered to build a WtE plant are: authorities, waste sector, energy sector and community
	Good information and communication are key elements together with a guaranteed participation process, strict regulation to control and monitor the emissions to the environment. Public audiences could help

7.2 BARRIERS FOR WtE IN ASIAN DEVELOPING COUNTRIES

Roshni Mary Sebastian, Dinesh Kumar, and Babu J. Alappat

Indian Institute of Technology, New Delhi, India

Waste incineration is undergoing significant growth globally. The number of active waste plants in the world is estimated to grow by >9% to 2700 by 2027 (Ecoprog, 2018). However, the global contribution of WtE sector to electricity generation is merely 0.4% (Massarutto, 2015). The energy from waste is renewable energy contributing to a considerable reduction in CO_2 emissions (Manders, 2009). One of the biggest challenges for the WtE industry in Asia is the thermal characteristics of the MSW feed. Besides, poor segregation operations in the developing countries make the generated MSW highly heterogeneous. With waste generation rates increasing, the complexity of waste segregation at source is only bound to get complex (Xin-Gang et al., 2016). The characteristics are also subject to seasonal variations, which can considerably affect the incinerability of MSW. Incineration has also been plagued by NIMBY syndrome, like any other waste management technique. Despite stringent emission standards being enforced, the technology is yet to garner public acceptance. Further, waste incineration is burdened by technical challenges as well. The improvement in the thermal characteristics and formulation of policies in the respective countries have made the scenario relatively conducive for waste incineration, of late. Some prominent challenges faced by waste incineration industry in Asia is as follows.

7.2.1 Technical Barriers in Asian Countries

7.2.1.1 Fluctuations in the Power Output

With the MSW feed being highly heterogeneous, the thermal characteristics like heat content, moisture content, volatile content, etc. are subject to wide variations. This, in turn, affects the power output from the facility.

7.2.1.2 Wear and Tear of Equipment

The production of HCl gas at a temperature > 350°C can cause corrosion of the combustion furnace which can reduce the efficiency as well as affect the economy of operation. With waste generated in Asia possessing high moisture content generally, the release of HCl, SO_2, and other acidic gases tend to be high which increases the chances of corrosion of the various units.

7.2.1.3 Inefficient Flue Gas Cleaning

Incineration of MSW results in the emissions of obnoxious pollutants which even after flue gas cleaning may not be completely eliminated. The advanced air pollution control units have, however, helped the incinerators adhere to the emission standards (Xin-Gang et al., 2016).

7.2.1.4 Ash Management

The residues from incinerators are primarily bottom ash and fly ash. While the former is not deemed to be hazardous in most countries, the latter is categorized as hazardous waste due to the content of heavy metals and dioxins. Reutilization of bottom ash is gaining momentum of late, with applications in road pavement construction, stone production, ceramics and glass industry, and even use as adsorbent. Pretreatment operations like washing or stabilization may also be carried out to remove any pollutants. In India, according to Solid Waste Management Rules, 2016, the bottom ash may be reused only if it has no toxic materials in it. The standards for fly ash disposal varies in different countries. It is generally directed to be disposed of in a hazardous waste landfill. However, more research needs to be done on the MSW fly ash reutilization potential (Zhang et al., 2015).

7.2.2 Economic Barriers in Asian Countries

7.2.2.1 Economic Feasibility of WtE Process in Asia

The economic feasibility of incineration to be a part of the sustainable solid waste management is a crucial aspect. A few studies on the economic aspect of waste incineration has been carried out in Thailand, China, Brazil, etc. (Leme et al., 2014; Massarutto, 2015; Menikpura et al., 2016; Murphy and McKeogh, 2004).

Massarutto (2015) categorizes the expenditure incurred on WtE plants as financial costs and external costs.

The former entails the installation and operational costs, which is influenced by the plant capacity, desired output from the plant, location of the plant, and the policy framework in the respective country. Two approaches are conventionally followed for quantifying the financial costs. The first approach is an engineering model involving experts to design the facility. While overnight costs for the plants are considered, location-specific expenses like land costs, compensation to the uprooted communities, etc. are not accounted for by this approach. These factors are later assumed and weighed in. Such desktop models work on numerous assumptions on operating time and maintenance intervals. However, the variability of the factors involved makes this approach ambiguous. The second approach uses data from operating WtE facilities. While this is a more realistic approach, the unavailability of reliable data due to the unwillingness of the plant operators to share the financial details makes it challenging.

The market price of the recovered heat and generated electricity offsets the installation and operational costs to some extent. In developed countries with the MSW possessing a high calorific value of >2500 kcal/kg, power generation can compensate for about 80%–90% of the total cost. Besides, these many countries add a subsidy or taxes to the electricity generated. The generated ash also finds a recycling value, which is dependent on national policies. For instance, the Italian market price for the ash is 20 euros/ton, in turn generating a revenue of 1.5 euro/ton of the MSW incinerated. Ash is treated as a hazardous by-product and its reuse potential is seldom explored. The management of ash increases the net cost by 7.5 euros/ton of the MSW incinerated. The revenue generated from materials recovered from the feed, viz., metals, glass, etc. has an insignificant impact on the net expenditure (Cossu and Masi, 2013; Dubois, 2013). It was identified that energy recovery can increase expenses by 30% (World Bank, 1999). This may be even higher for combined heat and energy recovery systems. The net investment cost can be brought down by nearly 30% if the plant is intended for heat recovery alone (Rand et al., 2000). Further, pollution abatement technologies, like air pollution control equipment accounts for nearly 60%–70% of the financial costs. The operating costs involve fixed expenses like administrative costs and salaries of employees. The number of skilled and unskilled laborers and the average labor costs in the respective country. The fixed operating costs are approximately 2% of the net investment costs.

External costs are composed of expenses due to air pollution control and management of incineration residues. The benefits due to the GHG emissions savings are also accounted for. However, this is influenced by the technology adopted in each plant and other location-specific factors. Furthermore, Massarutto (2015) reports that incineration of refuse-derived fuel (RDF) is less favorable than mass incineration due to the requirement for mechanical-biological treatment (MBT) facilities.

The financing of a waste incineration plant may come from a wide array of sources. This is tabulated in Table 7.2 (Rand et al., 2000).

A cost-benefit assessment study must justify higher treatment costs/ton of MSW prior to initiating the project. This is instrumental for assessing the feasibility of a WtE project. Waste transport costs, pollution potential of the process, the sustainability of energy generation, and policies and legislation should be considered for this assessment. If the results are negative, the project is deemed to be a "high-risk" project (Rand et al., 2000).

As discussed in preceding sections, the installation and operational costs are very high for WtE technologies. The installation of Shanghai Pudong Waste incineration plant incurred nearly 110 million US$.

Table 7.2 Primary Sources of Income for a WtE Facility

Source	Details
User fees	User fee collected from individual households and large industrial consumers contribute to net investment and operational costs. The fee from individual household is normally collected with taxes for municipal services and varies with the national policies
Gate fees	Incineration plants receive a tipping fee from large customers for disposing of their wastes in the plant. However, laws and policies should be stringent to prevent illegal dumping and ensure the waste gets delivered to the plant
Policies and incentives	To encourage the use of the incineration facility and prevent landfilling measures like subsiding incineration fees, increasing landfill gate fees, etc. may be adopted
Public subsidies	Subsidies include financial grants, low-interest loans for WtE plants, taxes
Sale of energy/heat	A steady demand for the energy generated from the plant can considerably offset the plant costs

Data from Rand, T., Haukohl, J., Marxen, U., 2000. Municipal Solid Waste Incineration, World Bank Technical Guidance Report. World Bank Technical Paper No. WTP462.

7.3 BARRIERS TO WtE IN AFRICA

William H.L. Stafford

Council for Scientific and Industrial Research, Stellenbosch, South Africa; Department of Industrial Engineering, University of Stellenbosch, Stellenbosch, South Africa

Africa's rapid growth in waste generation is largely driven by urbanization and increased wealth and consumption patterns. Waste to energy opportunities can fulfill two development objectives—the avoidance of pollution from effective waste treatment and the provision of energy with displacement of fossil fuels in the energy supply mix. With the increase in energy access and the demand for energy associated with the development and developing nations, such as Africa there are increasing opportunities for energy from MSW. However, a number of barriers prove to be obstacles to the widespread use of waste to energy in Africa.

Table 7.3 below presents a summary of the existing challenges related to both developed and developing countries.

7.3.1 Institutional Barriers and Market Failures

The cost of waste disposal and the societal impacts of waste are often not included in the cost accounting of waste to energy projects. Disposal to landfill is seen as the most cost-effective option for MSW in Africa, but this fails to include the loss of land value and urban development potential from landfills, the contamination of water bodies from landfill leachate and several other impacts. In many parts of Africa, landfills are burnt to reduce the volume of the waste at the landfill site which causes significant urban air pollution. These practices are seen as the option with the lowest cost to manage wastes

Table 7.3 Challenges for Developed and Developing Countries

Challenges	Developed Countries	Developing Countries
Population growth and increasing per capita consumption	Increase waste generation Improvement in waste management technology	Increase in waste generation Higher waste complexity Premature closure of disposal sites Larger number of waste pickers
Policy implementation	Stringent regulations Effective	Implementation of adapted policy Lack of enforcement Ineffective Illegal activities
Changes in waste composition	Introduction of suitable approaches such as incineration, composting, pyrolysis, etc.	Failure in existing waste management system Disturb the waste management facilities
Public participation	High due to high awareness Active participation: daily habits	Low due to low awareness Indifferent habits and refusal to change current habits Retaliate with illegal waste disposal
Informal recycling such as scavenger, etc.	Absence due to safety and hygiene factors	An important aspect that promotes recycling Unavoidable die to economic drivers Number will increase with nations' GDP Health concern
Recycling strategies	Practical, in line with governmental policy	Mainly white papers and not applicable for the implementation to the current waste management system
Existing waste management system	Promote 3Rs	Mainly serve to dispose waste

since the establishment or capital costs are low compared to effective waste management with waste to energy. However, this perceived lowest cost option fails to include the increasing social and environmental cost-burdens of landfilling and does not recognize the opportunity cost of alternative waste management using waste to energy. Waste to energy not only effectively manages burdens of waste disposal and treatment, but also delivers energy to society. Since energy is a domestic necessity and a factor of industrial production, increasing energy supply to developing countries is seen as a catalyst for development.

The lack of integration in the government institutions and effective public-private partnerships is also seen as an obstacle to waste to energy implementations. Waste treatment is seen as a role of government and local municipalities who treat it to meet discharge and disposal regulations without careful consideration of the costs of providing energy for treatment, and the opportunity of waste to energy is often forgone. In addition, energy supply monopolies and vested interest can limit the opportunities for integrating energy into waste management projects. For example, the entrenched dependency on coal as the main energy source for generating electricity and the existence of a single power utility operator in South Africa (Eskom) has made it challenging for independent WtE operators to access the grid and receive favorable tariffs for electricity supply. Electricity costs were in the past kept artificially low due to incomplete accounting for previous government infrastructure investments in coal power stations and the associated grid infrastructure. Furthermore, the ecological costs and social burdens of coal-fired power stations are not accounted for. These externalities include carbon dioxide and sulfur dioxide emissions, and carbon particulates that cause air pollution; as well as acid mine drainage causing water pollution.

From an institutional perspective, it was observed that waste management in the urban context is the domain of civil engineers, and constitutionally, is a function of local government (which sometimes relies on the water boards to execute). Energy technology sits between mechanical, electrical, and process engineers, and constitutionally, is a function of the national government. It is very challenging to work across these disciplines that have different mandates and objectives.

7.3.2 Technology Barriers

The principal technology consideration, which arose when exploring the barriers to energy from waste, was the characteristics of the wastes and the appropriate technologies. The main issues were identified as:

- Water content: Depending on the water content and organic composition, different technologies will be appropriate. Generally biodegradable wet organic waste is suited to anaerobic digestion for biogas production, while the thermal waste to energy options are more suited to dry biomass-requiring biomass <50% moisture and often preferably <20%. For dilute wastewaters, dewatering may be impractical due to the energy and cost but nutrient-rich wastewaters may be a source of water fertilizer for growing biomass.
- Waste composition: MSW composition of often heterogonous and somewhat variable, often and influenced by very local and context-specific factors—such as domestic consumption patterns and local industries. Many industrial wastes may contain components that are inhibitory to microbial growth or recalcitrant to degradation and may, therefore, require prior separation or pretreatment.
- Volume and seasonality: Volumes of wastes may be variable and differ in composition. Seasonality is particularly relevant to agricultural or agro-industry wastes.

While certain technologies are well established internationally, they have not been demonstrated in Africa, thus hampering large-scale implementation. The scalability and reliability of new technologies are not proven in the local context and there is a need for local demonstration of internationally established technologies. However, it should also be considered that technology designs are not always suited to the local context of a developing and many wastes to energy technologies are complex to build and implement and Africa can lack the human resource capacity for repairs and maintenance. There is a considerable lack of skills at all levels—from designing and implementation to operation—which limits the ability to build and operate energy from MSW projects in Africa.

7.3.3 Financial Risks and Barriers

Many waste to energy technologies are expensive both in terms of capital outlay and the skills/expertise required for maintenance. African municipalities or local private business often may not have the resources to pursue energy from waste projects, especially if long payback periods are encountered. In some cases, legislation surrounding the nature of such partnership contracts can limit the interest from private sector parties in pursuing such projects. For example, there may be legislation that limits the duration of contracting between public and private parties that places business risk on the security of waste supply for waste to energy projects. However, to date, the most common waste management option employed in Africa remains open dumping and uncontrolled burning of waste. Where waste policies and legislation are in place, implementation and enforcement lag behind, resulting in ineffective waste management. The private sector is often better positioned to provide efficient and effective waste management services, but policies are generally not supportive of partnerships. New public-private partnerships and financing structures, such as build-operate-transfer and build-own-operate-transfer, are needed to ensure that private sector is enabled to finance, design, construct, own, and operate a facility stated in the concession contract period with the public sector partner. This enables the private sector project developer to recover its investment and operating and maintenance expenses in the project in a defined time period while the public reaps the societal benefits.

Harnessing the potential of energy from MSW requires use of decision support and accounting that framework that considers the life cycle of the waste technology or solution, with the inclusion of several social and environmental costs and benefits. Using such approaches, the costs (capital and operational costs) and inputs (infrastructure, chemicals, and materials) and outputs (solid waste generation, water pollution, air pollution, and carbon emissions) of the various technologies appropriate to a given waste can be assessed and compared, and additional benefits, such as secondary products can also be taken into consideration. The net socioeconomic benefits of the energy from wastewater process includes the replacement of conventionally derived energy (i.e., coal-powered electricity), the generation of energy or fuel products and useful by-products as well as the loss of land value and avoiding impacts of landfilling, a reduction in pollution from wastes, and improved water resources management.

CHAPTER EIGHT

Concluding Remarks

Suani Teixeira Coelho[a]

Research Group on Bioenergy, Institute of Energy and Environment, University of São Paulo, São Paulo, Brazil

The adequate collection and disposal of municipal solid waste (MSW) remain a challenge in developing countries as a direct consequence of inadequate practices, which in turn produce negative environmental and social impacts. In industrialized countries most MSW are collected, reused, recycled, and, before being disposed in landfills, are recovered through waste-to-energy (WtE) systems. However, in developing countries, WtE technologies still face several barriers, in all aspects, and the synergies of WtE and basic sanitation are not yet well seen.

In this context, the main objective of this publication was to analyze the current situation of MSW collection and disposal, allowing the discussion of the perspectives of WtE in Latin America, Africa, and Asia.

As discussed in this book, the main MSW disposal is still landfills and often-in dumps. There are challenges mainly related to public awareness regarding the technologies, economic feasibility, and regulatory environment in all regions analyzed.

In LA, there is not enough public awareness of WtE systems as a desirable treatment option. There is, in general, lack of knowledge, on the part of public officials, the municipal and regional authorities, and the entities responsible for the management of MSW.

In addition, the LA region do not have a portfolio of companies that can manufacture equipment for incineration systems capable of handling the MSW, burning them in a controlled way, and fulfilling the environmental standards. Therefore, there is the need for imported equipment, making the investment much higher.

Important to note that in Brazil there are local manufacturers who have started developing small-scale thermal systems (MSW gasification) for small/medium municipalities, showing interesting perspectives.

In Asia, similar to LA, the not-in-my-backyard (NIMBY) syndrome, like any other waste management technique, has also plagued incineration. Despite stringent emission standards being enforced in the region, the technology is yet to garner public acceptance.

In all regions, installation and operational costs are very high for WtE technologies. Additionally, in Asia, there is the problem of pollutant emissions in the region. Even after

---

[a] Author acknowledges the important contribution from Prof. Jose Goldemberg, from University of São Paulo, Brazil.

flue gas cleaning, pollutants may not be completely eliminated. Advanced air pollution control units have however helped the incinerators adhere to the emission standards (Xin-Gang et al., 2016).

Besides the economic challenges, it is important to note that WtE technologies using thermochemical routes are a controversial issue in all regions, since they face some negative perception from the local society. Among them, incineration process in many cases is object of strong local rejection, firstly due to the fear of the pollutant emissions (mainly dioxins and furans). There is a significant lack of adequate information regarding the existing environmental legislation, as well as mandatory gas cleaning systems to fulfill the adequate environmental standards.

Another significant difficulty due to lack of information is related to the recycling process. There is often a strong belief that WtE processes is opposite to recycling. In many developing countries, there are several unskilled people working in dumps (exposed to huge health impacts), and there are also workers in cooperatives for manual recycling, such as in Brazilian municipalities. Therefore, there are concerns that WtE processes will oblige such people to lose their jobs. A widespread information program is necessary to explain that WtE processes do need previous recycling so, in regions where people do not perform their own waste recycling, the recycling process prior to WtE processes is mandatory.

On the other hand, waste collection services remain a real problem for the small municipalities, since incineration plants are economically feasible only for municipalities with at least 500,000–600,000 inhabitants.

In such cases, other options are necessary such as MSW gasification, which is possible to be installed in smaller municipalities.

In addition, an important aspect is the synergy between WtE and basic sanitation. The WtE technologies not only allow an increase in energy supply in the municipalities but also—and most important—contribute to solve the problem of adequate waste disposal.

There are few examples that illustrate this aspect. In Minas Gerais, Brazil, a small-scale WtE plant (1 MWe gasification plant using 55 t/d of MSW) is being built to solve the problem of inadequate waste disposal in a small municipality (Boa Esperança municipality, with 40,000 inhabitants, with the perspective to be expanded to the other municipalities nearby).

More important than the electricity generation (to be sold to the interlinked system) there is a solution to the environmental problem related to the inadequate MSW disposal (in a previous dump—with the liquid contaminant being discharged in to the nearby lake). However, when analyzing WtE economic feasibility, this analysis does not include the social, economic, and environmental benefits of the project when it avoids waste deposited in an open dump with all the negative impacts and costs.

In brief, benefits from WtE Technologies are a matter of fact. They include

(i) The positive aspects related to basic sanitation, since the process contributes to eliminate existing dumps and all negative impacts from them.

(ii) The contribution of WtE technologies to increase energy generation, not only in regions with adequate energy supply but also in regions where energy access is low or inexistent.

However, difficulties for the implementation of WtE technologies are huge. Challenges include the need for adequate information dissemination, for local capacity building, as well as the lack of funds for the investment and adequate regulations such as feed-in-tariffs and mandatory purchase, to allow the economic feasibility of the process. In general, the price of electricity generated by the WtE process is not competitive with the other sources of renewable energies. It would be interesting to analyze the possibility to guarantee a fixed subsidized price for the project at the time of implementation.

Conclusions show the high social and environmental benefits of WtE technologies for developing countries, as well as the contribution to increase energy offer in a sustainable way. Nevertheless, adequate policies and widespread information dissemination must be implemented to allow a sustainable MSW treatment, contributing to the development of the regions.

REFERENCES

ABRELPE, 2016. Panorama dos Resíduos Sólidos no Brasil 2016.

ABRELPE, 2017. Panorama dos Resíduos Sólidos no Brasil 2017.

ADB, 2009. Proposed Loan and Technical Assistance Municipal Waste to Energy Project (People's Republic of China). Asian Development Bank.

ADB, Asian Development Outlook 2017 Update—Sustaining Development Through Public-Private Partnership. https://www.adb.org/sites/default/files/publication/365701/ado2017-update.pdf. (Accessed 8 October 2017).

Adeyi, A.A., Adeyemi, A.M., 2017. Characterisation and leaching assessment of municipal solid wastes generated in Lagos and Ibadan, Nigeria. Nigerian J. Sci. 51 (1), 47–56.

AfDb, 2011. Africa in 50 Years' Time: The Road Towards Inclusive Growth. African Development Bank Group, African Development Bank, Tunis, Tunisia. Available at: https://www.afdb.org/fileadmin/uploads/afdb/Documents/Publications/Africa%20in%2050%20Years%20Time.pdf.

AGECC—The UN Secretary-General's Advisory Group on Energy and Climate Change, 2010. Energy for a Sustainable Future. Report and Recommendations. Available from: https://www.un.org/chinese/millenniumgoals/pdf/AGECCsummaryreport%5B1%5D.pdf.

Aguilar, S., 2015. La Promoción de Energías Renovables en Argentina: el caso GENREN. Puentes 15 (5). Análisis e Información sobre Comercio y Desarrollo Sostenible para América Latina ICTSD International Centre for Trade and Sustainable Development. https://www.ictsd.org/bridges-news/puentes/news/la-promoci%C3%B3n-de-energ%C3%ADas-renovables-en-argentina-el-caso-genren.

Alam, P., Ahmade, K., 2013. Impact of solid waste on health and the environment. Int. J. Sustain. Develop. Green Econom. V-2, I-1., 2, 2013.

AME/INEC, 2017. Estadística de Información Ambiental Económica en Gobiernos Autónomos Descentralizados Municipales - Gestión de Residuos Sólidos 2016. INEC, Quito, Ecuador. Available in: http://www.ecuadorencifras.gob.ec/documentos/web-inec/Encuestas_Ambientales/Gestion_Integral_de_Residuos_Solidos/2016/Documento%20tecnico%20Residuos%20solidos%202016%20F.pdf.

American Society for Testing and Materials (ASTM), 2013. ASTM Standard D-6700-01 (2013): Standard Practice for Use of Scrap Tire-Derived Fuel.

AMPC, 2017. Investigation into Modular Micro-Turbine Cogenerators & Organic Rankine Cycle Cogeneration Systems for Abattoirs—Final Report. Australian Meat Processor Corporation, Sydney, Australia, p. 39. Project Code 2016.1002.

ANEEL, 2018. Análise dos Planos de Universalização. Available in http://www.aneel.gov.br/planos-de-universalizacao. (Accessed 03 April 2018).

ANP (Agência Nacional do Petróleo, Gás Natural e Biocombustíveis), 2017. Resolução ANP no 685 de 29.06.2017. Brazil.

Aranda Uson, A., Lopez-Sabiron, A.M., Ferreira, G., Llera Sastresa, E., 2013. Uses of alternative fuels and raw materials in the cement industry as sustainable waste management options. Renew. Sust. Energ. Rev. 23, 242–260.

Aritono, H., 2017. Solid Waste to Energy, Department of Planning, Tanjung Pinang, Indonesia, Tanjung Pinang Municipality, UNESCAP, pp. 1–4. https://www.unescap.org/sites/default/files/CaseStudy_ID_Tanjungpinang_SolidWasteManagement_2017.pdf (Accessed 23 July 2019).

Arnold, J.E.M., Köhlin, G., Persson, R., Shepherd, G., 2003. Fuelwood Revisited: What Has Changed in the Last Decade? Center for International Forestry Research (CIFOR), Bogor, Indonesia.

ARS, 2012. Estrategia Nacional para la Gestión Integral de los Residuos Sólidos Urbanos. República Argentina, Diagnóstico de la GIRSU. Asociación Argentina para el Estudio de los Residuos Sólidos (ARS), Buenos Aires, Argentina.

Asamblea Constituyente, 2008. Constitución del Ecuador. Ciudad Alfaro.

Asamblea Nacional, 2010. Código Orgánico de Organización Territorial, Autonomía y Descentralización. R.O. No. 303.

Asamblea Nacional, 2017. Código Orgánico del Ambiente. R.O. No. 983.

Asian Development Bank (ADB), 2010. Guidelines for Estimating Asian Development Bank (ADB) Investments in Access to Energy Projects. Available from: http://www.asiandevbank.org/Documents/Clean-Energy/efa-brief.pdfS.

ASTM—American Society for Testing and Materials, 2013. Standard D-6700-01 (2013): Standard Practice for Use of Scrap Tire-Derived Fuel.

Ausubel, J., Marchetti, C., 1996. Electrical systems in retrospect and prospect. Daedalus. J. Am. Acad. Arts Sci. 139–169 (Summer 1996).

Ayeleru, O.O., Ntuli, F., Mbohwa, C., 2016. Proceedings of the World Conference on Engineering and Computer Science Vol (II), 19–21 October 2016, San Francisco, CA. http://www.iaeng.org/publication/WCECS2016/WCECS2016_pp625-629.pdf.

Bailis, R., Ezzati, M., Kammen, D.M., 2005. Mortality and greenhouse gas impacts of biomass and petroleum energy future in Africa. Science 308, 98–103.

Banga, M., 2011. Household knowledge, attitudes and practices in solid waste segregation and recycling: the case of urban Kampala. Zambia Soc. Sci. J. 2 (1), 4.

Banco Mundial, 2015. Diagnóstico de la Gestión Integral de Residuos Sólidos Urbanos en laArgentina. Recopilación, generación y análisis de datos – Recolección, barrido, transferencia, tratamiento y disposición final de Residuos Sólidos Urbanos. The World Bank.

Bast, E., Krishnaswamy, S., Mainhardt-Gibbs, H., Romine, T., 2011. Access to Energy for the Poor: The Clean Energy Option. Available from, http://actionaidusa.org/assets/pdfs/climate_change/Access_to_Energy_for_the_Poor_2011.pdfS.

Bazilian, M., Sagar, A., Detchon, R., Yumkella, K., 2010. More heat and light—viewpoint. Energy Policy 38, 5409–5412.

BCE, 2017a. Previsiones Macroeconómicas del Ecuador 2017–2020, Subgerencia de Programación y Regulación Monetaria y Financiera. Available in https://www.bce.fin.ec/index.php/component/k2/item/773-previsiones-macroecon%C3%B3micas. (Accessed 2 April 2018).

BCE, 2017b. Sistema de Información Macroeconómica - Estadísticas Macroeconómicas - Presentación Coyuntural - Marzo 2017 BCE. Available in: http://sintesis.bce.ec:8080/BOE/BI/logon/start.do?ivsLogonToken=bceqsappbo01:6400@2058546J8rEm61T5RRTVwh3ekKPs0z2058544JB3xSeIzJSiM3g3TiY7sIxp. (Accessed 2 May 2018).

BCE, 2018. Publicaciones Generales - Indicadores Macroeconómicos- Sector Externo - Balanza Comercial Anual 2017 BCE. Available in: http://sintesis.bce.ec:8080/BOE/BI/logon/start.do?ivsLogonToken=bceqsappbo01:6400@2210506JndImKNFw2QKa6nxETEhPZq2210504J1qDMpLg1Nu4nL5x5rfOKjF.

Bello, I.A., bin Ismael, M.N., Kabbashi, N.A., 2016. Solid waste management in Africa: a review. Int. J. Waste Resour. 6, 216.

Benhelal, E., Zahedi, G., Shamsaei, E., Bahadori, A., 2013. Global strategies and potentials to curb CO2 emissions in cement industry. J. Clean. Prod. 51, 142–161.

Bhatia, M., Angelou, N., 2015. Beyond connections. Energy access redefined. Technical Report 008/15. Energy Sector Management Assistance Program/The World Bank.

Blanco, G., Córdoba, V., Baldi, R., Fernández, M., Santalla, E., 2016. Outcomes of the clean development mechanism in Argentina. Am. J. Clim. Chang. 5, 431–447. http://www.scirp.org/journal/ajcc. ISSN Online: 2167-9509. ISSN Print: 2167-9495.

Madinah, N., Boerhannoeddin, A., Ariffin, R.N.B.R., 2014. Division solid waste generation and composition in Kampala Capital City Authority, Uganda: trends and management. IOSR J. Environ. Sci. Toxicol. Food Technol. 8 (10), 57–62.

Bond, T., Templeton, M.R., 2011. History and future of domestic biogas plants in the developing world. Energy Sustain. Dev. 15 (4), 347–354.

Borba, S., 2006. Análise de modelos de geração de gases em aterro sanitários: Estudo de caso. (Dissertação de Mestrado). Programa de Pós-Graduação de Engenharia Civil da Universidade Federal do Rio de Janeiro. Rio de Janeiro.

Border Environment Cooperation Commission (BECC), Cleanup of Juarez Scrap Tire Collection Center: A Successful Example of Binational Cooperation. http://www.becc.org/news/becc-news/cleanup-of-juarez-scrap-tire-collection-center-a-successful-example-of-binational-cooperation#.WufMzy7wbIU. (Accessed 30 April 2018).

Brasil, 2010. LEI No 12.305, DE 2 DE AGOSTO DE 2010. Brasília.

Braun, R., et al., 2010. Recent Development in Bio-energy Recovery Through Fermentation. Springer. https://doi.org/10.1007/978-3-642-04043-6_2.

Brew-Hammond, A., 2010. Energy access in Africa: challenges ahead. Energy Policy 38, 2291–2301.

Brew-Hammond, A., Kemausuor, F., 2009. Energy for all in Africa—to be or not to be. Curr. Opin. Environ. Sustain. 1, 83–89.

Bruce, N.G., Bates, E., Nguti, R., Gitonga, S., Kithinji, J., Doig, A., 2002. Reducing indoor air pollution through participatory development in rural Kenya. In: Proceedings of 9th International Conference on Indoor Air Quality and Climate, Monterey, CA, pp. 590–595.

Business Wire, 2018. https://oklahoman.com/article/feed/6501431/waste-to-energy-market-in-india-2018-researchandmarketscom.

Byrne, A., Gold, M., Turyasiima, D., Getkate, W., Niwagaba, C., Babu, M., Maiteki, J., Orwiny, M., Strande, L., 2015. Suitable Biowastes for Energy Recovery: Sludge to Energy Enterprises in Kampala (SEEK) Project. Eawag/Sandec, Dübendorf, Switzerland.

CAF, 2015a. Facilidad de Financiamiento climático Basado en Desempeño.

CAF, 2015b. Facilidad de Financiamiento Climático Basado en Desempeño CAF. Available in: http://ledslac.org/wp-content/uploads/2015/08/alejandromiranda_financiamientobasadoendesempeno.pdf. (Accessed 10 May 2018).

CAF, 2017. Loja, en Ecuador, apuesta por un desarrollo bajo en carbono y resiliente al cambio climático. CAF. Available in: https://www.caf.com/es/actualidad/noticias/2017/03/loja-en-ecuador-apuesta-por-un-desarrollo-bajo-en-carbono-y-resiliente-al-cambio-climatico/. (Accessed 12 May 2018).

CAF, 2018. Proyecto Huella de Ciudades. CAF. Available in: http://www.huelladeciudades.com/.

Cait Climate Data Explorer, 2017. Historical GHG Emissions. World Resources Institute, Washington, DC. Available in: http://cait.wri.org. (Accessed 4 May 2018).

Cammesa, 2018. Renewable inform. http://portalweb.cammesa.com/Documentos%20compartidos/Noticias/Mater/Informe%20Renovables%20FEB%202018.pdf. (Accessed February 2018).

Capstone Corporation, 2017. Available at: www.capstoneturbine.com/products. (Accessed 2 April 2017).

CDMX, 2017. Planta de termovalorización pone a la CDMX a la vanguardia en tecnología y materia ambiental. Retrieved April 19, 2018, from Ciudad de México: http://www.cdmx.gob.mx/comunicacion/nota/planta-de-termovalorizacion-pone-la-cdmx-la-vanguardia-en-tecnologia-y-materia-ambiental.

CEA, 2017–18. http://www.cea.nic.in/reports/annual/annualreports/annual_report-2018.pdf, 94 pp.

Cenbio, 2006. Centro Nacional de Referência em Biomassa. Primeiro Relatório de Atividades. Projeto Aproveitamento do Biogás Proveniente do Tratamento de Resíduos Sólidos Urbanos para Geração de Energia Elétrica e Iluminação a Gás. Meta 1/Fase 2 - Escolha e Caracterização do Aterro para a Implementação do Projeto. São Paulo, janeiro de.

CEMBUREAU, The European Cement Association. Main World Producer - The G20 Group. https://cembureau.eu/cement-101/key-facts-figures/. (Accessed 30 April 2018).

CENBIO/IEE/USP – Centro Nacional de Referência em Biomassa/Instituto de Energia e Ambiente/Universidade de São Paulo, 2006. Projeto GASEIFAMAZ. Available in http://cenbio.iee.usp.br/projetos/gaseifamaz/gaseifamaz.htm (Acessado em 2014).

Chanakya, H.N., Sharma, I., Ramachandra, T.V., 2009. Micro-scale anaerobic digestion of point source components of organic fraction of municipal solid waste. Waste Manag. 29 (4), 1306–1312.

Chen, X., Geng, Y., Fujita, T., 2010. An overview of municipal solid waste management in China. Waste Manag. 30, 716–724. https://doi.org/10.1016/j.wasman.2009.10.011.

Cheng, H., Hu, Y., 2010. Municipal solid waste (MSW) as a renewable source of energy: current and future practices in China. Bioresour. Technol. 101, 3816–3824. https://doi.org/10.1016/j.biortech.2010.01.040.

Chengula, A., Lucas, B.K., Mzula, A., 2015. Assessing the awareness, knowledge, attitude and practice of the community towards solid waste disposal and identifying the threats and extent of bacteria in the solid waste disposal sites in Morogoro Municipality in Tanzania. J. Biol. Agric. Healthc. 5 (3), 54–64.

China's National Climate Change Programme, 2007. Prepared under the Auspices of National Development and Reform Commission People's Republic of China, June. http://en.ndrc.gov.cn/newsrelease/200706/P020070604561191006823.pdf. (Accessed 22 July 2019).

China Statistical Yearbook, 2016. http://www.stats.gov.cn/tjsj/ndsj/2016/indexeh.htm. (Accessed 14 April 2018).

Cittadino, A., Igarzabal de Nistal, M.A., Zamorano, J., Ocello, J., Majul, M.V., D'hers, V., Ajhuacho, R., 2012. In: Wolkowicz (Ed.), Atlas de la Basura – Área Metropolitana de Buenos Aires, 151 pp.

Clarke, 2017. Available at: www.clarke-energy.com/biogas/. (Accessed 2 April 2017).

Clean Energy Finance Corporation, 2015. The Australian Bioenergy and Energy From Waste Market. https://www.cleanenergyfinancecorp.com.au/media/107567/the-australian-bioenergy-and-energy-from-waste-market-cefc-market-report.pdf.

Cloete, K., 2017. Ground-breaking waste-to-energy plant opens in Cape Town. Engineering News.

Coelho, S.T., Goldemberg, J., 2013. Energy access: lessons learned in Brazil and perspectives for replication in other developing countries. Energy Policy 61, 1088–1096. http://linkinghub.elsevier.com/retrieve/pii/S030142151300414X. (October 30, 2014).

Coelho, S.T., et al., 2015. Biomass residues as electricity generation source in low HDI regions of Brazil. In: Xi Latin-American Congress on Electricity Generation and Transmission—Clagtee 2015, p. 8.

Coelho, S., Cortez, C.L., Garcilasso, V.P., Escobar, J.F., Poveda, M.M., Coluna, N.M.E., 2016. Biomassa e Bioenergia. In: Energia e Sustentabilidade, vol. 1, first ed. Barueri, Manole, pp. 307–374.

Coelho, S.T., Sanches-Pereira, A., Castro, A.M., Calve, L.H., Infiesta, L.R., Miranda, L.H.T.G., 2017. Governança urbana e a recuperação energética dos resíduos sólidos. In: Governança mbiental global: diálogos com energia e meio ambiente, vol. 1, first ed. Leopoldianum, Sao Paulo, pp. 63–75.

Coelho, S.T. (Coord.), Garcilasso, V.P., Ferraz Jr., A.D.N., Santos, M.M., Joppert, C.L., 2018. Tecnologias de Produção e Uso de Biogás e Biometano. IEE-USP, São Paulo, ISBN: 978-85-86923-53-1.

Con la basura se proyecta generar energía en Urabá. El Colombiano. 19 de julio de. 2017.

CONAGUA, 2016. Comisión Nacional del Agua. Estadísticas del Agua en México, edición 2016. México. (www.gob.mx/conagua).

CONAPO, 2017. Datos de Proyecciones. Retrieved April 02, 2018, from Consejo Nacional de Población, http://www.conapo.gob.mx/es/CONAPO/Proyecciones_Datos.

Córdoba, V., Crozza, D., Santalla, E., 2008. Assessment, diagnosis and proposal actions for the improvement of environmental problematics and GHG mitigation linked to pig, cattle (feed lots and dairies) and poultry productions in Argentina. Evaluación, diagnóstico y propuestas de acción para la mejora de las problemáticas ambientales y mitigación de gases de efecto invernadero vinculadas a la producción porcina, bovina (feedlots y tambos) y avícola. 209 p. Inform for the World Bank BIRF through the Carbon Fund of Argentina. Contract No. 7145486, http://www.ambiente.gov.ar/?Idarticulo=6878.

Cossu, R., Masi, S., 2013. Re-thinking incentives and penalties: economic aspects of waste management in Italy. Waste Manag. 33, 2541–2547. https://doi.org/10.1016/j.wasman.2013.04.011.

Costa-Salas, Y., Sarache, W., Überwimmer, M., 2017. Fleet size optimization in the discarded tire collection process. Res. Transp. Bus. Manag. 24, 81–89.

Coviello, M.F., Ruchansky, B., 2017. Avances en materia de energías sostenibles en América Latina y el Caribe. Resultados del Marco de Seguimiento Mundial, informe de 2017. Documentos de Proyectos. CEPAL GTF.

Daurella, D.C., Foster, V., 2009. What can we learn from household surveys on inequalities in cooking fuels in sub-Saharan Africa? Available at: http://www.ecineq.org/ecineq_ba/papers/camos.pdf.

Davidson, O., Karekezi, S., 1992. A New, Environmentally Sound Energy Strategy for the Development of Sub-Saharan Africa. African Energy Policy Research Network.

Davidson, O., Sokona, Y., 2002. A New Sustainable Energy Path for African Development: Think Bigger, Act Faster. Energy and Development Research Centre (University of Cape Town) and the Environmental Development Action in the Third World (ENDA), Senegal. East African Community (EAC), 2008. Strategy for the development of regional refineries. Final Report.

DEA&DP, 2015. Greenest Municipality Competition Report 2015. Available at: https://www.westerncape.gov.za/eadp/files/atoms/files/GMC%20REPORT%202015_20160316.pdf.

de Queiroz Lamas, W., Fortes Palau, J.C., de Camargo, J.R., 2013. Waste materials co-processing in cement industry: ecological efficiency of waste reuse. Renew. Sust. Energ. Rev. 19, 200–207.

Department of Environmental Affairs (DEA), 2012. South Africa State of Waste. A report on the state of the environment. Final draft report. Department of Environmental Affairs, Pretoria.

Department of Environmental Affairs and Development Planning, Western Cape (DEADP), 2011. Status Quo Report: Integrated Waste Management Plan for the Western Cape Province. February 2011.

Department of Environmental Affairs and Tourism (DEAT), 2000. White Paper on Integrated Pollution and Waste Management for South Africa. A policy on Pollution Prevention, Waste Minimisation, Impact Management and Remediation. Government Gazette 20,978, Government Notice 227 of 17 March 2000.

DEWHA (Department of the Environment, Water, Heritage and the Arts), 2009. Employment in waste management and recycling. Report by Access Economics Pty Limited for DEWHA. Available in http://www.environment.gov.au/settlements/waste/publications/waste-and-recycling-employment. html. (Accessed 15 March 2018).

Dewees, P.A., 1989. The woodfuel crisis reconsidered: observations on the dynamics of abundance and scarcity. World Develop. 17 (8), 1159–1172.

Discourse Media, 2018. South East Asia Access to Energy Brief. Canada, Vancouver, BC, p. 3. www.discoursemedia.org. (Accessed 20 December 2018).

DCP – Documento de Concepção de Projeto/UNFCCC, 2012. Formulário do documento de concepção do projeto para atividades de projeto do MDL (F-CDM-PDD). Versão 04.1. Available in https://docs.google.com/viewer?a=v&pid=sites&srcid=ZGVmYXVsdGRvbWFpbnxjb25zdWx0aWFuxxxhne DozYTZhYjM5MjU3NzUyNmY. (Accessed May 2016).

Diaz-Chavez, R.A.D., Coelho, S.T., 2017. Waste governance and energy potential. In: lobal Environmental Issues: Law and Science, vol. 1, first ed. Leopoldianym University Press, Sao Paulo, pp. 109–124.

DNP, 2014. Bases del Plan Nacional de Desarrollo de Colombia para el periodo del 2014–2018. "Todos por un Nuevo País".

Documento CONPES 3874, 2016. Consejo Nacional de Política Económica y Social. Departamento Nacional de Planeación, Política Nacional Para la Gestión Integral de Residuos Sólidos, República de Colombia. Consulted on line in https://colaboracion.dnp.gov.co/CDT/Conpes/Econ%C3%B3micos/3874.pdf.

DOI/IEA, 2013. Energy-related carbon dioxide emissions. In: International Energy Outlook 2013. Washington, DC (Chapter 9).

Dong, J., Tang, Y., Nzihou, A., Chi, Y., Weiss-Hortala, E., Ni, M., Zhou, Z., 2018. Comparison of waste-to-energy technologies of gasification and incineration using life cycle assessment: case studies in Finland, France and China. J. Clean. Prod. 203, 287–300. https://doi.org/10.1016/j.jclepro.2018.08.139.

Downard, J., Singh, A., Bullard, R., Jayarathne, T., Rathnayake, C.M., Simmons, D.L., Wels, B.R., Spak, S.N., Peters, T., Beardsley, D., Stanier, C.O., Stone, E.A., 2015. Uncontrolled combustion of shredded tires in a landfill—part 1: characterization of gaseous and particulate emissions. Atmos. Environ. 104, 195–204.

Dube, R., Nandan, V., Dua, S., 2014. Waste incineration for urban India: valuable contribution to sustainable MSWM or inappropriate high–tech solution affecting livelihoods and public health? Technol. Manag. 17.

Dubois, M., 2013. Disparity in European taxation of combustible waste. Waste Manag. 33, 1575–1576. https://doi.org/10.1016/j.wasman.2013.05.001.

Dudek, J., Klimek, P., Kołodziejak, G., Niemczewska, J., Zaleska-Bartosz, J., 2010. Landfill Gas Energy Technologies. Oil and Gas Institute, National Research Institute, Cracow, Poland, p. 90.

EAI, 2013. India MSW to Energy—Status, Opportunities and Bottlenecks. Ministry of New and Renewable Energy. Available at: http://www.eai.in/ref/wp/indiamsw-to-energy.html. (Accessed 15 March 2018).

EBA (European Biogas Association), Available at: http://european-biogas.eu/2017/12/14/eba-statistical-report-2017-published-soon/. (Accessed 8 April 2017).

EBA (European Biogas Association), Statistical Report 2017. Available at: http://www.european-biogas.eu. (Accessed 8 April 2017).

EBA (European Biogas Association), 2017. Statistical Report 2017. Available at http://www.european-biogas.eu. (Accessed 4 August 2017).

Eberhard, A., Foster, V., Briceno-Garmendia, C., Ouedraogo, F., Camos, D., Shkaratan, M., 2008. Underpowered: The State of the Power Sector in SubSaharan Africa, African Infrastructure Country Diagnostic (AICD), Diagnostic Paper 6. World Bank, Washington, DC.

ECOCE, 2018. ECOCE. Retrieved April 26, 2018, from Cifras y estadíaticas, http://ecoce.mx/cifras.php.

Economic Commission for Africa (ECA), 2009. Challenges to Agricultural Development in Africa. Economic Report on Africa 2009. (Chapter 4). Available from: http://www.uneca.org/era2009/chap4.pdfS.

Ecoprog, 2018. https://www.ecoprog.com/publikationen/abfallwirtschaft/waste-to-energy.htm. (Accessed 16 April 2018).

EIA, 2013. India MSW to Energy-Status, Opportunities and Bottlenecks, Ministry of New and Renewable Energy. Available at: http://www.eai.in/ref/wp/indiamsw-to-energy.html. (Accessed 15 March 2018).

Eiceman, G.A., Clement, R.E., Karasek, F.W., 1979. Analysis of fly ash from municipal incinerators for trace organic compounds. Anal. Chem. 51, 2343–2350.

Ekere, W., Mugisha, J., Drake, L., 2009. Factors influencing waste separation and utilization among households in the Lake Victoria crescent, Uganda. Waste Manag. 29 (12), 3047–3051.

El reto de transformar la basura en energía. EL TIEMPO. 17 de agosto. 2015.

Elias, R.J., Victor, D.G., 2005. Energy Transitions in Developing Countries: A Review of Concepts and Literature. Working Paper no. 40, Program on Energy and Sustainable Development, Stanford University.

EMAC, 2015. Planta de biogas en Pichacay. EMAC. Available in: http://www.emac.gob.ec/?q=content/la-planta-de-biog%C3%A1s-en-pichacay-iniciar%C3%A1-su-funcionamiento. (Accessed 8 May 2018).

EMAC EP, 2013a. Aprovechamiento del biogás en el relleno sanitario de Pichacay.

EMAC EP, 2013b. Aprovechamiento del biogás en el relleno sanitario de PichacayEMAC. Available in: http://www.iner.gob.ec/wp-content/uploads/downloads/2013/05/EMAC_Aprovechamiento-del-biogas-en-el-relleno-sanitario-de-Pichacay_Galo_Vasquez.pdf. (Accessed 8 May 2018).

EMAC EP, 2015. Planta de biogas en Pichacay.

EMASEO, 2014. Índices de Gestión EMASEO Enero 2014. EMASEO, Quito, Ecuador. Available in: http://www.emaseo.gob.ec/documentos/2014/boletin_indices_gestion_enero2014.pdf. (Accessed 8 May 2018).

ENGIRSU, 2005. Estrategia Nacional Para La Gestión Integral de Residuos Sólidos Urbanos: ENGIRSU. República Argentina, Ministerio de Salud y Ambiente, Secretaría de Ambiente y Desarrollo Sustentable. Septiembre 2005.

ENGIRSU, 2016. Estrategia Nacional Para La Gestión Integral de Residuos Sólidos Urbanos: ENGIRSU. Mapas Críticos Gestión de Residuos. República Argentina, Ministerio de Salud y Ambiente, Secretaría de Ambiente y Desarrollo Sustentable. Marzo.

En marcha proyecto que convertirá basura de Medellín en energía. El Colombiano. 12 de diciembre de. 2016.

Energy Information Administration (EIA), 2011. International Energy Statistics. Available from: http://www.eia.gov/ipdbproject/IEDIndex3.cfmS.

ENT/MAE/URC/GEF, 2013a. Evaluación de Necesidades Tecnológicas para la Generación de Energía a partir de Residuos Sólidos Urbanos - Mitigación - Sector Energía - Ecuador Country Report. Ministerio del Ambiente MAE, Quito, Ecuador.

ENT/MAE/URC/GEF, 2013b. Evaluación de Necesidades Tecnológicas para la Generación de Energía a partir de Residuos Sólidos Urbanos - Mitigación - Sector Energía - Ecuador Country Report. Ministerio del Ambiente MAE, Quito, Ecuador. Available in: http://unfccc.int/ttclear/misc_/StaticFiles/gnwoerk_static/TNR_CRE/e9067c6e3b97459989b2196f12155ad5/ff118ec7a6df4f2d9dc4c5a6dd169473.pdf. (Accessed 11 May 2013).

European Council (EC), 2008. Directive 2008/98/EC of the European Parliament and of the Council of 19November 2008 on waste and repealing certain directives.Off. J.Eur.Union. 22/11/2008, L.312/3-30.

European Tyre & Rubber Manufacters' Association (ETRMA), 2017. Annual report. http://www.etrma.org/uploads/Modules/Documentsmanager/20170905—etrma-annual-report-2016-17—final.pdf. (Accessed 30 April 2018).

European Union (EU), 2015. Draft EN 16723-2 Natural Gas and Biomethane for Use in Transport and Biomethane for Injection in the Natural Gas Network—Part. 2 Automotive Fuel Specifications. CEN, Brussels.

European Union (EU), 2016. EN 16723-1 Natural Gas and Biomethane for Use in Transport and Biomethane for Injection in the Natural Gas Network—Part 1: Specifications for Biomethane for Injection in the Natural Gas Network. CEN, Brussels.

Eurostat, 2017. Municipal Waste Statistics. http://ec.europa.eu/eurostat/statistics-explained/index.php/Municipal_waste_statistics. (Accessed August 2017).

The European Cement Association (CEMBUREAU), 2018. Main world producer – the G20 group. https://cembureau.eu/cement-101/key-facts-figures/. (Accessed 30 April 2018).

Eurostat, 2018. Municipal Waste Treatment, EU-28, (Kg per Capita).

Fachverband Biogas, 2016. Biowaste to Biogas. Freising. [Online]. Available: http://www.biowaste-to-biogas.com/.

Fagundes, L.D., Santos Amorim, E., da Silva Lima, R., 2017. Action research in reverse logistics for end-of-life tire recycling. Syst. Pract. Action Res. 30, 553–568.

Fall, L., 2010. Achieving Energy Efficiency in Africa: What Are the Priorities, the Best Practices and the Policy Measures? World Energy Council, London, UK.

FAO, 1993. Selected Wood Energy Data. Food and Agriculture Organization of the United Nations. Available at: http://www.fao.org/3/w7744e/w7744e07.htm.

FAO, 2016. Global Forest Products: Facts and Figure. Food and Agriculture Organisation. Available at: http://www.fao.org/3/i7034en/i7034en.pdf.

Federación Interamericana del Cemento (FICEM), 2018. Cement Review Report – Ninth ed. – International Cement Review. http://www.ficem.org/asociados-latinoamerica/Estadisticas-Colombia.pdf. (Accessed 30 April 2018).

Feresu, S.B. (Ed.), 2013. Zimbabwe Environment Outlook: Our Environment, Everybody's Responsibility. The Ministry of Environment and Natural Resources Management, Government of the Republic of Zimbabwe, Harare, Zimbabwe.

Ferguson, H., 2012. Briquette Businesses in Uganda. The Potential for Briquette Enterprises to Address the Sustainability of the Ugandan Biomass Fuel Market. London, UK.

Former Ministry of Health and Welfare of Japan, 1996. The Interim Report Given by a Research Group for Dioxins and the Present Correspondence. dated June 28 (in Japanese), Office of Life Chemical Safety Measures, Environmental Health Bureau, Tokyo.

Forrest, M., 2014. Overview of the world rubber recycling market. In: Forrest, M. (Ed.), Recycling and Re-use of Waste Rubber. (Chapter 3). Available at https://www.smithersrapra.com/SmithersRapra/media/Sample-Chapters/Recyclingand-Re-use-of-Waste-Rubber.pdf. (Accessed 30 April 2018).

Fouquet, R., 2008. Heat, Power and Light: Revolutions in Energy Services. Edward Elgar, Cheltenham, UK.

Frandsen, T.Q., Rodhe, L., Baky, A., Edström, M., Sipilä, I.K., Petersen, S.L., Tybirk, K., 2011. Best available technologies for pig manure biogas plants in the Baltic Sea region. In: Baltic Sea 2020, Stockholm.

Fundação Estadual do Meio Ambiente (FEAM), 2010. Estudo do estado da arte e análise de viabilidade técnica, econômica e ambiental da implantação de uma usina de tratamento térmico de resíduos sólidos urbanos com geração de energia elétrica no estado de Minas Gerais: Relatório 1. 2. ed. Belo Horizonte – Minas Gerais.

GAIA, 2003. Global Alliance for Incinerator Alternatives. Waste Incineration: A Dying Technology. Pg no. 32. Available at: http://www.no-burn.org/wp-content/uploads/Waste-Incineration-A-Dying-Technology.pdf.

Gates Foundation, 2012. Solid Waste Management: Characterisation on the African Continent. Characteristing Waste in Five Cities. Final Report.

Gaye, A., 2007. Access to Energy and Human Development. Human Development Report 2007/2008. Fighting Climate Change: Human Solidarity in a Divided World. Available from: http://hdr.undp.org/en/reports/global/hdr2007-8/papers/.

Gaye_Amie, Grubler, A., 1991. Diffusion: long-term patterns and discontinuities. Technol. Forecast. Soc. Chang. 39 (1–2), 159–180 (ISSN 0040-1625).

GBIO, 2015. Biomass Residues as Energy Source to Improve Energy Access and Local Economic Activity in Low HDI Regions of Brazil and Colombia (BREA). São Paulo. Available at: www.iee.usp.br/gbio.

GEF/UNIDO, 2014. Projeto GEF – Cuba. Available in: http://www.cubaheadlines.com/2014/10/08/38930/opened_seminar_about_biomass_gasification_in_havana.html; http://gasifiers.bioenergylists.org/cubaenergia.

GIZ, 2010. Guia Prático do Biogás Geração e Utilização. Tradução Eng. Ftal. Marcos de Miranda Zattar. Editor Fachagentur Nachwachsende Rohstoffe e. V. (FNR) 5ª. Ed.

GIZ, 2011. https://www.resource-recovery.net/sites/default/files/giz_ghana_kenia_mg-mt_14.09.2015_-_english.pdf.

GIZ-EnRes, 2016. Potencial para la valorización energética de residuos. GIZ, Ciudad de México, México.

GIZ-EnRes, 2019. Proyectos de aprovechamiento energético a partir de residuos urbanos en México. GIZ, Ciudad de México, México, 32 pp. Retrieved 24 March 2018. Available at https://www.bivica.org/files/5407_giz-enres.pdf.

Global Energy Assessment (GEA), 2012. Towards a Sustainable Energy Future. Cambridge University Press/International Institute for Applied Systems Analysis, Cambridge UK and New York, NY, USA/Laxenburg, Austria. Available at http://www.globalenergyassessment.org.

Gobierno del Ecuador, 2017. Código Orgánico del Ambiente.

GoK, 1999. The Republic of Kenya. Environmental Management and Coordination Act 1999.

GoK, 2010. The Constitution of Kenya 2010.

GoK, 2012. The County Government Act Nairobi, Kenya 2012.

Goldenberg, J., 2000. Rural energy in developing countries. In: World Energy Assessment: Energy and the Challenge of Sustainability. UNDP, New York.

Gouveia, N., Prado, R.R., 2010. Riscos à saúde em áreas próximas a aterros de resíduos sólidos urbanos. Revista Saúde Pública. 44(5), 859–866.

Government of India, 2011. National Implementation Plan: Stockholm Convention on Persistent Organic Pollutants. 104 pp. https://www.indiawaterportal.org/sites/indiawaterportal.org/files/National_Implementation_Plan_Stockholm_Convention_on_Persistent_Organic_Pollutants_MoEF_2011.pdf. (Accessed 23 July 2019).

Government of Japan, 2012. Dioxins.

Government of Zimbabwe (GoZ), 2009. Zimbabwe National Environmental Policy and Strategies. Government of the Republic of Zimbabwe, Ministry of Environment and Natural Resources Management, Harare.

Government of Zimbabwe (GoZ), 2013. National Energy Policy. Ministry of Energy and Power Development, Harare.

Gray, K.R., 2003. Multilateral environmental agreements in Africa: efforts and problems in implementation. Int. Environ. Agreements: Pol. Law Econ. 3, 97–135.

Gu, B., Jiang, S., Wang, H., Wang, Z., Jia, R., Yang, J., He, S., Cheng, R., 2017. Characterization, quantification and management of China's municipal solid waste in spatiotemporal distributions: a review. Waste Manag. 61, 67–77. https://doi.org/10.1016/j.wasman.2016.11.039.

Gujba, H., Thorn, S., Rai, K., Mulugetta, Y., Sokona, Y., 2012. Financing energy access in the context of low carbon development in Africa. In: African Climate Policy Centre Working Paper Series, No. 14. United Nations Economic Commission for Africa, Addis Ababa, Ethiopia.

Gullett, B.K., Dunn, J.E., Raghunathan, K., 2000. Effect of co-firing coal on formation of polychlorinated dibenzo-p-dioxins and dibenzofurans during waste combustion. Environ. Sci. Technol. 34, 282–290.

Hakawati, R., Smyth, B.M., Mccullough, G., De Rosa, F., Rooney, D., 2017. What is the most energy efficient route for biogas utilization: Heat, electricity or transport? Appl. Energy 206, 1076–1087.

Harare City Council, 2016. Request for Expression of Interest for Establishment of a Waste to Energy Plant at Pomona Landfill Site Through a Private Public Partnership. October 2016.

Haregu, T.N., Ziraba, A.K., Mberu, B., 2016. Integration of Solid Waste Management Policies in Kenya: Analysis of Coherence, Gaps and Overlaps. https://www.nema.go.ke/index.php?option=com_content&view=article&id=1&Itemid=136.

Henry, R.K., Yongsheng, Z., Jun, D., 2006. Municipal solid waste management challenges in developing countries—Kenyan case study. Waste Manag. 26 (1), 92–100.

Hershkowitz, A., Salerni, E., 1989. Municipal solid waste incineration in Japan *. Environ. Impact Assess. Rev. 9, 257–278.

Hitachi Zosen, 2014. http://www.mofa.go.jp/region/latin/fealac/pdfs/4-9_jase.pdf. (Accessed 19 April 2018).

Hitachi Zosen, 2018. http://www.hitachizosen.co.jp/english/pickup/pickup002.html. (Accessed 13 April 2018).

Hoornweg, D., Bhada-Tata, P., 2012a. What a Waste: A Global Review of Solid Waste Management. Urban Development Series Knowledge Papers. World Bank, Washington, DC.

Hoornweg, D., Bhada-Tata, P., 2012b. A Global Review of Solid Waste Management. Urban Development Series Knowledge Papers. World Bank, pp. 1–116.

Hoornweg, D., Bhada-Tata, P., Kennedy, C., 2015. Peak waste: when is it likely to occur? J. Ind. Ecol. 19 (1), 117–128. https://doi.org/10.1111/jiec.12165.

Hot Times, Waste-to-Energy Plants Burn Bright in China's Cities. Burning Up: Rapid Growth in China's Waste-to-Energy Plants. Consulted on line in https://www.newsecuritybeat.org/2017/11/default-post/.

How to make Waste-to-Wealth a reality. The Economic Times, Indian Times. January 14, 2018. https://inventariogei.ambiente.gob.ar/files/2doBUR%20ARGENTINA.pdf.

https://thecitywasteproject.files.wordpress.com/2013/03/solid_waste_management_in_the_worlds-cities.pdf.

Hrbek, J., 2015. Thermal gasification based hybrid systems. IEA Bioenergy Task 33 Special Project. ISBN 978-1-910154-52-6.

Huang, Q.F., Wang, Q., Dong, L., Xi, B., 2006. The current situation of solid waste management in China. J. Mater. Cycles Waste Manag. 8, 63–69.

Huang, Q., Chi, Y., Themelis, N.J., 2013. A rapidly emerging WTE technology: circulating fluid bed combustion. In: Proceedings of International Thermal Treatment Technologies (IT3), San Antonio, TX, October. Available at https://mafiadoc.com/a-rapidly-emerging-wte-technology-circulating-fluid-bed-_5a1ea5c01723dd7ab88b9d2b.html.

Hughes, J., 2013. Small Scale Environmental Solutions—A New Market Opportunity. June 21, http://www.pennog.com/small-scale-environmental-solutions-a-new-market-opportunity/.

IBGE, 2013. Projeções Da População: Brasil E Unidades Da Federação.

IBGE, 2017. Contas Regionais 2015: Queda No PIB Atinge Todas as Unidades Da Federação Pela Primeira Vez Na Série.

ICLEI, 2017. Gestão adequada de resíduos nas cidades pode ajudar Brasil a cumprir meta de Paris. http://sams.iclei.org/novidades/noticias/arquivo-de-noticias/2017/lancamento-novo-seeg.html. (Accessed 15 March 2018).

IDEAM, 2015. Inventario nacional de gases de efecto invernadero GEI Colombia. Consulted on line in: http://documentacion.ideam.gov.co/openbiblio/bvirtual/023421/cartilla_INGEI.pdf.

IEA, 2006. Energy for Cooking in Developing Countries. International Energy Agency (IEA), World Energy Outlook. Available at: https://www.iea.org/publications/freepublications/publication/cooking.pdf.

IEA, 2007. International Energy Agency. Retrieved from Biomass for Power Generation and CHP. https://www.iea.org/publications/freepublications/publication/essentials3.pdf.

IEA, 2009. Energy Balance. International Energy Agency. Available at http://iea.org/stats/prodresult.asp?PRODUCT=Balances.

IEA, 2014. https://www.instituteforenergyresearch.org/fossil-fuels/coal/ieas-world-energy-outlook-2014/#_edn, 3 pp.

IEEJ, 2016. https://www.ief.org/_resources/files/snippets/ieej/ieej_outlook2016__7007_rv_for_ief.pdf, p. 10 (Accessed 20 July 2019).

IFENG, "Opinions on Promoting the Service in Distributed Power Parallel to Grid" issued by the state grid. http://finance.ifeng.com/roll/20130228/7715935.shtml. (Accessed 3 May 2013).

IIASA, 2005. Global Energy Assessment – Energy, Poverty, and Development. International Institute for Applied Systems Analysis, Laxenburg, Austria.

IIASA, 2012. Global Energy Assessment (GEA): Toward a Sustainable Future. Laxenburg.

IISc – Indian Institute of Sciente, Bangalore, India, 2010. Biomass Gasification. Available in http://cgpl.iisc.ernet.in/site/Technologies/BiomassGasification/tabid/68/Default.aspx (Acessado em 2014).

Imam, A., Mohammed, B., Wilson, D.C., Cheeseman, C.R., 2008. Solid waste management in Abudja, Nigeria. Waste Manag. 28 (2), 468–472.

INEC, 2010. Resultados del Censo 2010. INEC, Quito, Ecuador. Available in: http://www.inec.gov.ec/cpv/index.php?option=com_wrapper&view=wrapper&Itemid=49&lang=es. (Accessed 23 March 2012).

INEC, 2012. Proyección de la Población Nacional 2010–2050. Available in: http://www.ecuadorencifras.gob.ec/documentos/web-inec/Poblacion_y_Demografia/Proyecciones_Poblacionales/presentacion.pdf. (Accessed 2 May 2018).

INEC, 2017a. Índice de Precios al Consumidor - Canastas - Histórico. INEC. Available in: http://www.ecuadorencifras.gob.ec/canasta/. (Accessed 2 April 2018).

INEC, 2017b. Estadísticas sociales - Encuesta Nacional de Empleo, Desempleo y Subempleo -ENEMDU 2017. INEC. Available in: http://www.ecuadorencifras.gob.ec/enemdu-2017/. (Accessed 2 May 2018).

INECC, 2018. Investigaciones 2018–2013 en materia de mitigación del cambio climático. Retrieved April 16, 2018, from Instituto Nacional de Ecología y Cambio Climático: https://www.gob.mx/inecc/documentos/investigaciones-2018-2013-en-materia-de-mitigacion-del-cambio-climatico.

INEGI (n.d.). Cuéntame…población. Retrieved April 20, 2018, from Instituto Nacional de Geografía y Estadística: http://cuentame.inegi.org.mx/poblacion/densidad.aspx?tema=P.

Indian Census, 2011. Important Facts PDF Download–Census 2011 pdf. https://www.studydhaba.com/indian-census-2011-important-facts-pdf-download/ (Accessed 30 June 2019).

Infiesta, L., 2015. Gaseificação de resíduos sólidos urbanos (RSU) no Vale do Paranapanema: Projeto CIVAP. Monografia (Especialização em Energias Renováveis)–Escola Politécnica, Universidade de São Paulo, São Paulo.

Inovageo, 2017. Geomembrana de PEAD. http://inovageo.eng.br/produtos/geomembrana/?gclid=EAIaIQobChMI2IzEtIam2gIVFQSRCh0_3QtTEAAYASAAEgI-NPD_BwE/. (Accessed 13 November 2017).

Instituto Brasileiro do Meio Ambiente e dos Recursos Naturais Renováveis (IBAMA), 2018. Ministério do Meio Ambiente. Relatório Pneumáticos 2018: Resolução Conama nº 416/09 2018 (ano-base 2017). https://www.ibama.gov.br/phocadownload/pneus/relatoriopneumaticos/ibama-relatorio-pneumaticos-2018.pdf. (Accessed 29 November 2018).

Instituto Nacional de Estadísticas y Censo (INDEC), 2010. https://www.indec.gov.ar/nivel4_default.asp?id_tema_1=2&id_tema_2=41&id_tema_3=135.

Instituto Nacional de Estadísticas y Censo (INDEC), 2018. https://www.indec.gov.ar/nivel2_default.asp?id_tema=2&seccion=P.

International Energy Agency (IEA), 2010. World Energy Outlook 2010. Available from: http://www.iea.org/Textbase/npsum/weo2010sum.pdfS.

International Energy Agency (IEA), 2011a. Energy for All: Financing Energy Access for the Poor. Special Early Report on the World Energy Outlook 2011. Available from: http://www.iea.org/papers/2011/weo2011_energy_for_all.pdfS.

International Energy Agency (IEA), 2011b. Statistics and Balances. Available from: http://www.iea.org/stats/index.aspS.

INTI, 2016. Final Inform OT N 615000021. National Survey of Anaerobic Digestion Plants With Electric and/or Thermal Exploitation. National Institute of Industry Technology, Renewable Energy Program, Special Projects Management.

IPCC, 1996. Revised IPCC Guidelines for National Greenhouse Gas Inventories, 1996.

IPCC, 2013. Working Group I contribution to the IPCC 5th assessment report "Climate Change 2013: The Physical Science Basis"—approved summary for policymakers. Available at: http://www.ipcc.ch/report/ar5/wg1/#.UlRN1tJLM33.

Japan Environment Ministry, 2015. Municipal Waste Incineration Technology, Safe and Sound Waste Incineration and High-Efficiency Power Generation. Waste Recovery Commission, Japan Environment Ministry, Tokyo, Japan.

Japan Environmental Facilities Manufacturers Association (JEFMA), 2018. http://www.jefma.or.jp/englishpage_f.htm. (Accessed 15 April 2019).

Jiang, J., Lou, A., Ng, S., Luobu, C., Ji, D., 2009. The current municipal solid waste management situation in Tibet. Waste Manag. 29, 1186–1191.

Jimenez, R., Yépez García, A., 2016. Composition and Sensitivity of Residential Energy Consumption. Inter-American Development Bank Infrastructure and Energy Sector-Energy Division.

Jiménez, R., Yépez-García, A., 2017. Understanding the drivers of household energy spending, micro evidence for Latin America. IDB Working Paper Series, 805. IDB.

Kadas, M.B., Fraker, R., Martella Jr., R.R., 2014. Central and South America overview: emerging trends in Latin America. (Chapter 20). In: Grosko, J.B. (Ed.), International Environmental Law. The Practitioner's Guide to the Laws of the Planet. American Bar Association.

Kammen, D.M., 2006. Bioenergy in developing countries: experiences and prospects. In: Hazell, P., Pachauri, R.K. (Eds.), Bioenergy and Agriculture: Promises and Challenges. IFPRI, Washington, DC.

Karak, T., Bhagat, R.M., Bhattacharyya, P., 2012. Municipal solid waste generation, composition, and management: the world scenario. Crit. Rev. Environ. Sci. Technol. 42, 1509–1630. https://doi.org/10.1080/10643389.2011.569871.

Kathirvale, S., Muhd Yunus, M.N., Sopian, K., Samsuddin, A.H., 2004. Energy potential from municipal solid waste in Malaysia. Renew. Energy 29 (4), 559–567.

Katusiimeh, M.W., Mol, A.P.J., Burger, K., 2012. The operations and effectiveness of public and private provision of solid waste collection services in Kampala. Habitat Int. 36 (2), 247–252.

Kazungu, R.K., 2010. Improving governance for sustainable waste management in Nairobi. In: Proceedings of the 46th ISOCARP Congress, 2010.

KCC, 2006. Solid Waste Management Strategy Report. Kampala City Council, Kampala, Uganda.

KCCA, 2017. Kampala Waste Treatment and Disposal PPP: Project Teaser. [Online]. Available from: https://www.kcca.go.ug/uDocs/kampala-waste-treatment-and-disposal-ppp.pdf. (Accessed 15 March 2018).

Kebede, E., Kagochi, J., Jolly, C.M., 2010. Energy consumption and economic development in Sub-Sahara Africa. Energy Econ. 32 (3), 532–537.

Khan, I.U., Othman, M.H.D., Hashim, H., Matsuura, T., Ismail, A.F., Rezaei-Dashtarzhandi, M., Azelee, I., 2017. Biogas as a renewable energy fuel—a review of biogas upgrading, utilization and storage. Energy Convers. Manag. 150, 277–294.

Khajuria, A., Yamamoto, U., Morioka, T., 2010. Estimation of municipal solid waste generation and landfill area in Asian developing countries. J. Environ. Biol. 31 (5), 649–654.

Kinobe, J.R., Niwagaba, C.B., Gebresenbet, G., Komakech, A.J., Vinnerås, B., 2015. Mapping out the solid waste generation and collection models: the case of Kampala City. J. Air Waste Manage. Assoc. 65 (2), 197–205.

Komakech, A.J., Banadda, N.E., Kinobe, J.R., Kasisira, L., Sundberg, C., Gebresenbet, G., Vinnerås, B., 2014. Characterization of municipal waste in Kampala, Uganda. J. Air Waste Manage. Assoc. 64 (3), 340–348.

Komen, K., Mtembu, N., van Niekerk, M.A., 2016. The role of socio-economic actors, seasonality and geographic differences on household waste generation and composition in the City of Tshwane. In: Proceedings of the 23rd WasteCon Conference, 17–21 October 2016, Emperors Palace, Johannesburg, South Africa.

Krishan, M.P.P.L., 2012. GULU BIO-ENERGY: providing sustainable energy in the northern war-torn region of Uganda. Available at: https://english.rvo.nl/sites/default/files/2013/12/Biogas%20from%20digestion%20-%20Uganda%20-%20MPPL%20Renewable%20Energy%20Pvt.Ltd_.pdf.

Kubanza, N.S., Simatele, D., 2016. Social and environmental injustices in solid waste management in sub-Saharan Africa: a study of Kinshasa, the Democratic Republic of Congo. Local Environ. 21 (7), 866–882. https://doi.org/10.1080/13549839.2015.1038985.

Kumar, S., Bhattacharyya, J.K., Vaidya, A.N., Chakrabarti, T., Devotta, S., Akolkar, A.B., 2009. Assessment of the status of municipal solid waste management in metro cities in India: An insight. Waste Manag. 29, 883–895.

Kumar, S., Smith, S., Fowler, G., Velis, C., Kumar, J., Arya, S., Rena, Kumar, R., Cheeseman, C., 2017. Challenges and opportunities associated with waste management in India. R. Soc. Open Sci. 4 (3), 160764. Published online 2017 Mar 22, https://doi.org/10.1098/rsos.160764.

Lata, K., Rajeshwari, K.V., Pant, D.C., 2001. TEAM process: conceptualization on efforts to meet the challenges of vegetable market waste management problem. Bioenergy News 5 (1), 21–23.

Le Courtois, A., 2012. Municipal solid waste: turning a problem into a resource. Priv. Sect. Develop. (15). Available online at http://www.ccacoalition.org/en/resources/municipal-solid-waste-turning-problem-resource.

Leach, G., 1992. The energy transition. Energy Policy 20 (2), 116–123.

Lederer, J., Ongatai, A., Odeda, D., Rashid, H., Otim, S., Nabaasa, M., 2015. The generation of stake-holder's knowledge for solid waste management planning through action research: a case study from Busia, Uganda. Habitat Int. 50, 99–109.

Leme, M.M.V., Rocha, M.H., Lora, E.E.S., Venturini, O.J., Lopes, B.M., Ferreira, C.H., 2014. Techno-economic analysis and environmental impact assessment of energy recovery from Municipal Solid Waste (MSW) in Brazil. Resour. Conserv. Recycl. 87, 8–20. https://doi.org/10.1016/j.resconrec.2014.03.003.

LEY 1715 DE, 2014. Diario Oficial No. 49.150 de 13 de mayo de 2014. Consulted on line in http://servicios.minminas.gov.co/compilacionnormativa/docs/pdf/ley_1715_2014.pdf.

Liu, Z., Liu, Z., Li, X., 2006. Status and prospect of the application of municipal solid waste incineration in China. Appl. Therm. Eng. 26, 1193–1197. https://doi.org/10.1016/j.applthermaleng.2005.07.036.

Lohri, C.R., 2012. Feasibility assessment tool for urban anaerobic digestion in developing countries. In: Environmental Sciences (Major in Environmental Technology). Wageningen University, Netherlands.   https://www.eawag.ch/fileadmin/Domain1/Abteilungen/sandec/publikationen/SWM/Anaerobic_Digestion/Lohri_2012.pdf. (Accessed 22 July 2019).

Lohri, C.R., Diener, S., Zabaleta, I., Mertenat, A., Zurbrügg, C., 2017. Treatment technologies for urban solid biowaste to create value products: a review with focus on low- and middle-income settings. Rev. Environ. Sci. Biotechnol. 16 (1), 81–130.

Louw, K., Conradie, B., Howells, M., Dekenah, M., 2008. Determinants of electricity demand for newly electrified low-income African household. Energy Policy 36 (8), 2771–3232.

Lwasa, S., Kadilo, G., 2010. Participatory action research, strengthening institutional capacity and governance: confronting the urban challenge in Kampala. Commonwealth J. Local Gov, https://doi.org/10.5130/cjlg.v0i5.1467.

Lwasa, S., Nyakaana, J., Senyendo, H., 2007. Population, Urban Development and the Environment in Uganda: The Case of Kampala City and Its Environs. Makerere University, Kampala.

Ma, J., Hipel, K.W., 2016. Exploring social dimensions of municipal solid waste management around the globe—a systematic literature review. Waste Manag. 56, 3–12.

Machin, E.B., Pedroso, D.T., de Carvalho Jr., J.A., 2017. Energetic valorization of waste tires. Renew. Sust. Energ. Rev. 68, 306–315.

MAE, 2014. Manual Básico de Aprovechamiento Energético de Residuos Agropecuarios. Quito.

MAE, (Ed.), 2017a. Tercera Comunicación Nacional del Ecuador a la Convención Marco de las Naciones Unidas sobre el Cambio Climático, first ed. MAE, Quito, Ecuador.

MAE, 2017b. Ficha informativa de proyecto K009 MAE - Programa Nacional para la Gestión Integral de Desechos Sólidos. MAE-PNGIDS.

MAE, 2017c. Ficha informativa de proyecto K009 MAE - Programa Nacional para la Gestión Integral de Desechos Sólidos (MAE-PNGIDS) MAE/PNGIDS. Available in: http://www.ambiente.gob.ec/wp-content/uploads/downloads/2017/12/PNGIDS-NOVIEMBRE-2017.pdf. (Accessed 8 May 2018).

Makara, S., 2009. Decentralisation and urban governance in Kampala. (PhD Thesis). University of the Witwatersrand, Johannesburg, South Africa.

Malimbwi, R., Chidumayo, E.N., Zahabu, E., Kingazi, S., Misana, S., Luoga, E., Nduwamungu, J., 2010. Woodfuel. In: Chidumayo, E.N., Gumbo, D.J. (Eds.), The Dry Forests and Woodlands of Africa: Managing for Products and Services. Earthscan, London, UK, pp. 155–177.

Malmsheimer, R.W., Heffernan, P., Brink, S., Crandall, D., Deneke, F., Galik, C., Gee, E., Helms, J.A., McClure, N., Mortimer, M., Ruddell, S., Smith, M., Stewart, J., 2008. Forest management solutions for mitigating climate change in the United States. Journal of Forest. 106 (3), 115–117. https://doi.org/10.1093/jof/106.3.115.

Manders, I., 2009. The renewable energy contribution from waste across Europe. In: ISWA Dakofa Conference, December 3.

Mani, S., Singh, S., 2016. Sustainable municipal solid waste management in India: a policy agenda. Procedia Environ. Sci. 35, 150–157.

Mark, E., 2012. Assessing the use of power generation technologies in Uganda: a case study of Jinja Municipality. Diss KTH industrial Engineering and Management.

Martínez, J.D., Lapuerta, M., García-Contreras, R., Murillo, R., García, T., 2013a. Fuel properties of tire pyrolysis liquid and its blends with diesel fuel. Energy Fuel 27, 3296–3305.

Martínez, J.D., Puy, N., Murillo, R., García, T., Navarro, M.V., Mastral, A.M., 2013b. Waste tyre pyrolysis—a review. Renew. Sust. Energ. Rev. 23, 179–213.

Martínez, J.D., Murillo, R., García, T., Veses, A., 2013c. Demonstration of the waste tire pyrolysis process on pilot scale in a continuous auger reactor. J. Hazard. Mater. 261, 637–645.

Martínez, J.D., Cardona-Uribe, N., Murillo, R., García, T., López, J.M., 2019. Carbon black recovery from waste tire pyrolysis by demineralization: production and application in rubber compounding. Waste Manag. 85, 574–584.

Masera, O.R., Saatkamp, B.D., Kammen, D.M., 2000. From linear fuel switching to multiple cooking strategies: a critique and alternative to the energy ladder model. World Develop. 28 (12), 2083–2103.

Massarutto, A., 2015. Economic aspects of thermal treatment of solid waste in a sustainable WM system. Waste Manag. 37, 45–57. https://doi.org/10.1016/j.wasman.2014.08.024.

Mawatha, S., Lanka, S., 2015. Market assessment of RRR business models—Kampala City Report. Sri Lanka.

Mbuligwe, S.E., 2012. Solid Waste Management Assessment Within Urban Settings in Burundi, Rwanda and Tanzania. LVWATSAN Programme Phase II Report UN-Habitat.

Mbuligwe, S.E., Kassenga, G.R., Kaseva, M.E., Chaggu, E.J., 2002. Potential and constraints of composting solid waste in developing countries: findings from a pilot study in Dar es Salaam, Tanzania. Resour. Conserv. Recycl. 36, 45–59.

McAllister, J., 2015. Factors Influencing Solid Waste Management in the Developing World. All Graduate Plan B and other Reports, 528. Available online at https://digitalcommons.usu.edu/gradreports/528.

McBean, E., 2008. Siloxanes in biogases from landfills and wastewater digesters. Can. J. Civ. Eng. 35, 431–436.

MCIDADES.SNSA, 2012. Sistema Nacional de Informações Sobre Saneamento: Diagnóstico Do Manejo de Resíduos Sólidos Urbanos – 2010. Brasilia.

McPeak, J.G., 2003. Fuelwood Gathering and Use in Northern Kenya: Implications for Food Aid and Local Environments. USAID Global Livestock CRSP Research Brief 03-01-PARIMA.

MEER, 2016. Plan Maestro de Electricidad 2016–2025. Ecuador.

MEER, 2017. Plan Maestro de Electricidad 2016–2025. Ecuador.

Meidiana, C., Gamse, T., 2010. Development of waste management practices in Indonesia. Eur. J. Sci. Res. 40 (2), 199–210.

Menezes, R.A.A., Gerlach, J.L., Menezes, M.A., 2000. Estágio Atual da Incineração no Brasil. In: Seminário Nacional de Resíduos Sólidos e Limpeza Pública, 7, Curitiba, 3 a 7 de abril de 2000. Anais…ABL – Associação Brasileira de Limpeza Publica, Curitiba.

Menikpura, S.N.M., Sang-Arun, J., Bengtsson, M., 2016. Assessment of environmental and economic performance of Waste-to-Energy facilities in Thai cities. Renew. Energy 86, 576–584. https://doi.org/10.1016/j.renene.2015.08.054.

Menya, E., Alokore, Y., Ebangu, B.O., 2013. Biogas as an alternative to fuelwood for a household in Uleppi sub-county in Uganda. Agric. Eng. Int. CIGR J. 15 (1), 50–58.

MEWC, 2014. Zimbabwe Integrated Solid Waste Management Plan. Ministry of Environment, Water and Climate, Government of the Republic of Zimbabwe, Harare.

MEWC, 2015a. Zimbabwe's National Climate Change Response Strategy. Ministry of Environment, Water and Climate, Government of the Republic of Zimbabwe, Harare.

MEWC, 2015b. Zimbabwe's Intended Nationally Determined Contribution (INDC) Submitted to the United Nations Framework Convention on Climate Change (UNFCCC). Available at http://www4.unfccc.int/submissions/INDC/Published%20Documents/Zimbabwe/1/Zimbabwe%20Intended%20Nationally%20Determined%20Contribution%202015.pdf.

MEWC, 2016. Zimbabwe's Third National Communication to UNFCCC. Ministry of Environment, Water and Climate, Government of the Republic of Zimbabwe, Harare.

MICSE, 2016. Balance Energético Nacional.

MIDUVI, 2015. Informe Nacional del Ecuador - Subsecretaría de Hábitat y Asentamientos Humanos 2015 - SHAH Tercera Conferencia de las Naciones Unidas sobre la Vivienda y el Desarrollo Urbano Sostenible

Habitat III. Quito, Ecuador. Available in: https://www.habitatyvivienda.gob.ec/wp-content/uploads/downloads/2017/05/Informe-Pais-Ecuador-Enero-2016_vf.pdf. (Accessed 8 May 2018).

MIDUVI, 2016. Informe Nacional del Ecuador - Subsecretaría de Hábitat y Asentamientos Humanos 2015 - SHAH Tercera Conferencia de las Naciones Unidas sobre la Vivienda y el Desarrollo Urbano Sostenible Habitat III. Quito, Ecuador.

Min Ambiente – Gobierno de Colombia – Noticias, 2016. A 2018 Colombia tendrá una tasa de reciclaje del 20%. Consulted on line in: http://www.minambiente.gov.co/index.php/noticias/2291-a-2018-colombia-tendra-una-tasa-de-reciclaje-del-20.

Ministerio de Hidrocarburos, 2018. Rendición de Cuentas 2017. Available in: http://www.hidrocarburos.gob.ec/wp-content/uploads/2018/03/Informe-de-Rendicion-de-Cuentas-2017.pdf. (Accessed 28 March 2018).

Ministerio de Vivienda, Ciudad y Territorio. DECRETO 1077 DE 2015. Diario Oficial No. 49.523 de 26 de mayo de 2015. Por medio del cual se expide el Decreto Único Reglamentario del Sector Vivienda, Ciudad y Territorio.

Ministerio de Vivienda, Ciudad y Territorio. República de Colombia. Comisión de Regulación de Agua Potable y Saneamiento Básico. RESOLUCIÓN CRA 720 DE 2015. "Por la cual se establece el régimen de regulación tarifaria al que deben someterse las personas prestadoras del servicio público de aseo que atiendan en municipios de más de 5.000 suscriptores en áreas urbanas, la metodología que deben utilizar para el cálculo de las tarifas del servicio público de aseo y se dictan otras disposiciones". Consulted on line: http://www.metropol.gov.co/Residuos/Documents/.

Ministerio del Ambiente, 2014a. Acuerdo Ministerial No. 061, R.O. E.E. No. 316.

Ministerio del Ambiente, 2014b. Acuerdo Ministerial No. 061.

Ministerio del Ambiente, 2014c. Manual Básico - Aprovechamiento Energético de Residuos Agropecuarios - Enfocado a la Mitigación del Cambio Climático, 1ra ed. MAE, Quito, Ecuador.

Ministerio del Ambiente, 2015. Acuerdo Ministerial No. 061, Registro Oficial Edición Especial No. 316.

Ministry of Construction, 2007. Municipal Solid Waste Management Approaches, Beijing, China. http://www.gov.cn/ziliao/flfg/2007-06/05/content_636413.htm (in Chinese).

Ministry of Economy, Trade, and Industry, 2018. Feed-in Tariff Scheme in Japan. Available at: http://www.meti.go.jp/english/policy/energy_environment/renewable/pdf/summary201207.pdf. (Accessed 14 April 2018).

Ministry of Environment, Forest and Climate Change, 2016. Solid Waste Management Rules. Government of India. http://www.moef.nic.in/content/so-1357e-08-04-2016-solid-waste-management-rules-2016?theme=moef_blue. (Accessed 17 April 2018).

Ministry of the Environment, 2012. Solid Waste Management and Recycling Technology of Japan—Towards a Sustainable Society.

Ministry of the Environment, 2015. Solid Waste Management and Recycling Technology of Japan—Towards a Sustainable Society.

Ministry of Water and Environment (MWE), 2014. Water and Environment Sector Performance Report 2014. Ministry of Water and Environment, Kampala, Uganda.

Miyingo, E.W., 2012. Medical solid waste incineration effectiveness and energy recovery possibilities: case study of medical solid waste in Kampala Health units—Uganda. Diss Makerere University.

MMA, 2018. Política Nacional de Resíduos Sólidos. https://www.mma.gov.br/pol%C3%ADtica-de-res%C3%ADduos-s%C3%B3lidos. (Accessed 29 August 2019).

MME/EPE, 2017. Plano Decenal de Expansão de Energia 2026. Brasília, Rio de Janeiro.

MNRE, 2013. https://mnre.gov.in/file-manager/annual-report/2012-2013/EN/index.html. (Accessed 22 July 2019).

Modi, V., McDade, S., Lallement, D., Saghir, J., 2005. Energy Services for the Millennium Development Goals. UN Millienium Project of the International Bank for Reconstruction and Development/The World Bank and the United Nations Development Programme. Available from: http://www.unmillenniumproject.org/documents/MP_Energy_Low_Res.pdfS.

Moghadam, M.R.A., Mokhtarani, N., Mokhtarani, B., 2009. Country report: municipal solid waste management in Rasht City. Iran. Waste Manag. 29, 485–489.

Mohee, R., Simelane, T., 2015. Future Directions of Municipal Solid Waste Management in Africa. Africa Institute of South Africa, Pretoria, South Africa.

MoHUA, 2018. Waste To Wealth: A Ready Reckoner for Selection of Technologies for Management of Municipal Solid Waste.

Mokyr, J., 1990. The Lever of Riches. Oxford University Press, New York.

Mongkolnchaiarunya, J., 2005. Promoting a community-based solid-waste management initiative in local government: Yala municipality, Thailand. Habitat Int. 29, 27–40.

MoUHA Newsletter, 2017.

Moulot, J., 2005. Unlocking rural energy access for poverty reduction in Africa. In: Proceedings of the 15th Congress of the Union of Producers, Transporters and Distributors of Electric Power in Africa (UPDEA) Accra, Ghana. United Nations Economic Commission for Africa.

Mugagga, F., 2006. The Public–Private Sector Approach To Municipal Solid Waste Management: How Does It Work In Makindye Division, Kampala District, Uganda?. Master of Philosophy in Development Studies Specializing in Geography Thesis, Department of Geography, Norwegian University of Science and Technology (NTNU).

Mugalu, M., 2015. City Abattoir Waste Turns to Electricity. The Observer Kampala, Uganda.

Mulugetta, Y., Doig, A., Jackson, T., Khennas, S., Dunnett, S., Rai, K., 2005. Energy for Rural Livelihoods: A Framework for Sustainable Decision Making. IT Publications, London.

Mundial, Banco, 2015. Diagnóstico de la Gestión Integral de Residuos Sólidos Urbanos en laArgentina. Recopilación, generación y análisis de datos – Recolección, barrido, transferencia,tratamiento y disposición final de Residuos Sólidos UrbanosThe World Bank.

Murphy, J.D., McKeogh, E., 2004. Technical, economic and environmental analysis of energy production from municipal solid waste. Renew. Energy 29, 1043–1057. https://doi.org/10.1016/j.renene.2003.12.002.

Muswere, G.K., RodicWiersma, L., 2004. Municipal Solid Waste Management in Greater Harare, Zimbabwe.

Muylaert, M., Ambram, R., Campos, C., Montez, E., Oliveira, L., 2000. Consumo de energia e aquecimento do planeta – Análise do mecanismo de desenvolvimento limpo (MDL) do Protocolo de Quioto – Estudo de Caso. Coppe, Rio de Janeiro, 247 pp.

Nahman, A., 2011. Pricing landfill externalities: emissions and disamenity costs in Cape Town, South Africa. Waste Manag. 31, 2046–2056.

Naidoo, R., 2017. New biogas plant opens in the cape. Infrastructure News.

National Bureau of Statistics of China, 2015. Annual Data. http://www.stats.gov.cn/tjsj/ndsj/2015/indexch.htm. (Accessed 16 April 2018).

NCPD, 2013. Kenya Population Situation Analysis.

NEMA, 2006. State of the Environment Report for Uganda 2006/2007. National Environment Management Authority, Kampala, Uganda.

NEMA, 2014. National State of the Environment Report for Uganda 2014. "Harnessing our Environment as Infrastructure for Sustainable Livelihood & Development" National Environment Management Authority, Kampala, Uganda.

NEPAD, 2000. New Partnership for Africa's Development. http://www.nepad.org/.

Ngwenya, P., Jonsson, L., 2001. Industrial Waste management in Zimbabwe.

Nie, Y., 2008. Development and prospects of municipal solid waste (MSW) incineration in China. Front. Environ. Sci. Eng. China 2, 1–7. https://doi.org/10.1007/s11783-008-0028-6.

Niemczewska, J., 2012. Characteristics of Utilization of Biogas Technology. vol. 68. Nafta-Gaz, pp. 293–297.

Njenga, M., Karanja, N., Prain, G., Malii, J., Munyao, P., Gathuru, K., Mwasi, B., 2009. Community-Based Energy Briquette Production from Urban Organic Waste at Kahawa Soweto Informal Settlement. Nairobi. Urban Harvest Working Paper Series, No. 5. ISSN 1811-1440.

NNFCC, 2010. A Detailed Economic Assessment of Anaerobic Digestion Technology and Its Suitability to UK Farming and Waste Systems. A project funded by DECC and managed by the NNFCC. The Andersons Centre. Project No. NNFCC 08-006.

Noel, C., 2010. Solid waste workers and livelihood strategies in Greater Port-au-Prince, Haiti. Waste Manag. 30, 1138–1148.

NSSO, 2016. http://mospi.nic.in/sites/default/files/publication_reports/mospi_Annual_Report_2016-17.pdf, 89 pp.

NSWMP, 2011. National solid waste management Programme, Egypt. Main Report, December 2011. MoLD/EEAAA/KfW.

Nuss, P., Bringezu, S., Gardner, K., 2012. Waste to materials. The long-term option. In: Karagiannidis, A. (Ed.), Waste to Energy. Opportunities and Challenges for Developing and Transition Economies. Springer, New York, pp. 1–26.

Nussbaumer, P., Bazilian, M., Modi, V., Yumkella, K.K., 2012. Measuring energy poverty: focusing on what matters. Renew. Sust. Energ. Rev. 16 (2012), 231–243.

Nwufo, C.C., 2010. Legal framework for the regulation of waste in Nigeria. Afr. Res. Rev. 4 (2), 491–501.

Nyakaana, J.B., 1997. Solid waste management in urban centers: The Case of Kampala City-Uganda. East Afr. Geogr. Rev. 19 (1), 33–43.

Nzeadibe, T.C., 2015. Moving up the hierarchy: involving the informal sector to increase recycling rates in Nigerian cities. In: Mohee and Simelane (Ed.), Future Directions of Municipal Solid Waste Management in Africa. Africa Institute of South Africa. 268 pages.

Nzila, C., Dewulf, J., Spanjers, H., Kiriamiti, H., van Langenhove, H., 2010. Biowaste energy potential in Kenya. Renew. Energy 35 (12), 2698–2704.

O'Connor, P.A., 2010. Energy Transitions. The Pardee Papers: no 12. The Frederick S. Pardee Center for the Study of the Longer-Range Future. Boston University. Available from: http://www.bu.edu/pardee/files/2010/11/12-PP-Nov2010.pdfS.

Oboirien, B.O., North, B.C., 2019. A review of waste tyre gasification. J. Environ. Chem. Eng. 5, 5169–5178.

OECD, 1974. Declaration on Environmental Policy. 14 November. C/M(74) 26/FINAL. Consulted on line in: https://legalinstruments.oecd.org/Instruments/.

OECD, 2018. https://stats.oecd.org/Index.aspx?DataSetCode=MUNW. (Accessed 18 April 2018).

Oelofse, S.H.H., Godfrey, L., 2008. Defining waste in South Africa: moving beyond the age of waste. S. Afr. J. Sci. 104, 242–246. July/August 2008.

Oelose, S.H.H., Mouton, C., 2014. The impacts of regulation on business in the waste sector. Evidence from the Western Cape. In: Proceedings of WasteCon 2014. Lord Charles Hotel Somerset West, Cape Town, 6–10 October 2014.

Office of Auditor General, 2010. Value for Money Audit Report on Solid Waste Management in Kampala. Office of Auditor General (OAG), Republic of Uganda. [Online]. Available from: http://www.oag.go.ug/wp-content/uploads/2016/07/SOLID-WASTE.pdf. (Accessed 14 March 2018).

Ojok, J., Koech, M., Tole, M., OkotOkumu, J., 2013. Rate and quantities of household solid waste generated in Kampala City, Uganda. Sci. J. Environ. Eng. Res. 2013.

Okello, C., Pindozzi, S., Faugno, S., Boccia, L., 2013. Development of bioenergy technologies in Uganda: a review of progress. Renew. Sust. Energ. Rev. 18, 55–63.

Okot-Okumu, J., 2012. Solid waste management in African cities—East Africa. In: Rebellon, L.F.M. (Ed.), Waste Management—An Integrated Vision. InTech, Rijeka (Chapter 1).

Okot-Okumu, J., 2015. Solid waste management in Uganda: challenges and options. In: Mohee, R., Simelane, T. (Eds.), Future Directions of Municipal Solid Waste Management in Africa. Africa Institute of South Africa, South Africa.

Okot-Okumu, J., Nyenje, R., 2011. Municipal solid waste management under decentralisation in Uganda. Habitat Int. 35 (4), 537–543.

OLADE, 2017. Anuario de 2017 Estadísticas Energéticas. OLADE, Quito, Ecuador.

OLADE/CEPAL/GTZ, 2003. Energía y desarrollo sustentable en ALyC: Guía para la formulación de políticas energéticas, Santiago de Chile, Cepal.

Olie, K., Vermeulen, P.L., Hutzinger, O., 1977. Chlorodibenzo-p-dioxins and chlorodibenzofurans are trace components of fly ash and flue gas of some municipal incinerators in the Netherlands. Chemosphere 8, 455–459.

Oliveira, L.B., Rosa, L.P., 2003. Brazilian waste potential: energy, environmental, social and economic benefits. Energy Policy 31 (14), 1481–1491.

Osepeense, 2015. O Portal de Notícias de São Sepé e Região. Conheça a primeira usina do RS a gerar energia a partir do lixo. Available in http://osepeense.com/conheca-a-primeira-usina-do-rs-a-gerar-energia-a-partir-do-lixo/. (Accessed March 2018).

Osibanjo, O., 2002. Hazardous Wastes. Invited Seminar Lecture Presented to the Parliamentary Committee on Environment and Poverty Alleviation. Tanzania Parliament, Dodoma, Tanzania.

Otage, S., 2016. Uganda: NWSC to Begin Generating Electricity, Biogas in 2017. Daily Monitor Kampala, Uganda.

Oteng-Ababio, M., Arguella, J.E.M., Gabbay, O., 2013. Solid waste management in African Cities: sorting the facts from the fads in Accra, Ghana. Habitat Int. 39, 96–104.

Othiang, E.O., 2018. Uganda to Construct Five Waste-to-Energy Plants. Construction Review Online Nairobi, Kenya.

Oyoo, R., Leemans, R., Mol, A.P.J., 2014. Comparison of environmental performance for different waste management scenarios in East Africa: the case of Kampala City, Uganda. Habitat Int. 44, 349–357.

Pachauri, S., Spreng, D., 2003. Energy Use and Energy Access in Relation to Poverty. CEPE Working Paper Series 03-25 CEPE Center for Energy Policy and Economics, ETH Zurich.

Packaging, S.A., 2015. Design for Recycling for Paper and Packing in South Africa. Available at: http://www.packagingsa.co.za/wp-content/uploads/2015/09/Packaging-SA-Recyclability-by-Design-2015.pdf.

Park, J., Díaz Posada, N., Mejía Dugand, S., 2018. Challenges in implementing the extended producer responsibility in an emerging economy: the end-of-life tire management in Colombia. J. Clean. Prod. 189, 754–762.

Pasang, H., Moore, G.A., Sitorus, G., 2007. Neighbourhood-based waste management: a solution for solid waste problems in Jakarta, Indonesia. Waste Manag. 27, 1924–1938.

Patrizio, P., Leduc, S., Chinese, D., Dotzauer, E., Kraxner, F., 2015. Biomethane as transport fuel—a comparison with other biogas utilization pathways in northern Italy. Appl. Energy 157, 25–34.

Persson, M., Jönsson, O., Wellinger, A., 2006. Biogas Upgrading to Vehicle Fuel Standards and Grid Injection. IEA Bioenergy Task 37.

Petrobras, 2014. Refinaria gera energia com biogás do Aterro de Gramacho. Available in http://www.petrobras.com.br/fatos-e-dados/refinaria-gera-energia-com-biogas-do-aterro-de-gramacho.htm. (Accessed May 2016).

Phiri, A., Mwanza, B., 2013. Design of a waste management model using integrated solid waste management: a case of Bulawayo City council. Int. J. Water Resour. Environ. Eng. 5 (2), 110–118. http://www.academicjournals.org/IJWREE.

Plan para generar energía con residuos sigue en pie (this is for Bogotá). Portafolio septiembre 03 de. 2014.

Planning Commission, 2014. Report of the Task Force on Waste to Energy. Vol. I. Planning Commission, New Delhi.

PNUD, 2016. Panorama general. Informe sobre Desarrollo Humano 2016. Desarrollo humano para todos. Programa de la Naciones Unidas para el Desarrollo. One United Nations Plaza, New York, NY, Estados Unidos.

PNUD, 2017. Informe Nacional sobre Desarrollo Humano 2017. Información para el desarrollo sostenible: Argentina y la Agenda 2030/dirigido por Gabriela Catterberg y Ruben Mercado; edición literaria a cargo de Sociopúblico; con prólogo de René Mauricio Valdés, 1.a ed. Programa Naciones Unidas para el Desarrollo - PNUD, Buenos Aires.

Posada, E., 2012. Reflexiones sobre el presente y el futuro de la ingeniería de proyectos. Dyna, año 79, Edicion Especial Medellín, pp. 14–16. ISSN 0012-7353.

Posada, E., 2017. The culture of innovation and sustainable development: challenges for engineering. Int. J. Develop. Res. 7 (12), 17655–17660.

Posada, E., 2018. Strategic analysis of alternatives for waste management. In: Kumar, S. (Ed.), Waste Management (Chapter 2). ISBN 978-953-7619-84-8.

Presentations of Keynote Speakers at Recycle 2018 at IIT GUWAHATI from Feb 22–24, 2018. https://drive.google.com/file/d/1c15R9jGmYLv618wZbeJYVK5OyrEBZOE2/view?usp=sharing.

Priestley, J., 1790. Experiments and Observations on Different Kinds of Air. vol. 1. Birmingham. 206 pp. (Quoted by Tietjen, 1975).

Rago, Y.P., Mohee, R., Surropp, D., 2012. A review of thermochemical technologies of waste biomass to biofuel and energy in developing countries. In: Leal Filho, W., Surroo, D. (Eds.), The Nexus: Energy, Environment and Climate Change. Springer, Switzerland, pp. 127–144.

Ramachandra, T.V., Bachamanda, S., 2007. Environmental audit of municipal solid waste management. Int. J. Environ. Technol. Manag. 7 (3/4), 369–391.

Ramachandran, V., 2008. Power and Roads for Africa. Centre for Global Development. Available from, http://www.cgdev.org/files/16557_file_Africa_Brief_web.pdfS.

Rand, T., Haukohl, J., Marxen, U., 2000. Municipal Solid Waste Incineration. World Bank Technical Guidance Report. World Bank Technical Paper No. WTP462.

Recalde, M.Y., 2016. The different paths for renewable energies in Latin American Countries: the relevance of the enabling frameworks and the design of instruments. WIREs Energy Environ. 5 (3), 305–326.

Recalde, M., 2017. La inversión en energías renovables en Argentina. Rev. Econ. Inst. 19 (36), 231–254.

Recalde, M.Y., Ramos-M, J., 2012. Going beyond energy intensity to understand the energy metabolism of nations: the case of Argentina. Energy 37 (1), 122–132.

Recalde, M., Guzowski, C., Zilio, M., 2014. Are modern economies following a sustainable energy consumption path? Energy Sustain. Dev. 19, 151–161.

Recalde, M.Y., Bouille, D., Girardin, L.O., 2015. Limitaciones para el desarrollo de energías renovables en Argentina: el rol de las condiciones de marco desde una perspectiva histórica. Rev. Probl. Desarr. 183 (46), 89–115. octubre-diciembre 2015.

Reed, T.B., Graboski, M., Markson, M., 1982. The SERI high pressure oxygen gasifier. Report SERI/TP-234-1455R, February, Solar Energy Research Institute, Golden, Colorado.

Reister, D., 1987. The link between energy and GDP in developing countries. Energy 12 (6), 427–433.

Rellenos sanitarios darán electricidad a Colombia. PORTAFOLIO junio 12 de. 2013.

REN21, 2011. Renewables 2011 Global Status Report. Renewable Energy Policy Network for the 21st Century. Available from: http://www.ren21.net/Portals/97/documents/GSR/REN21_GSR2011.pdfS.

REN21, 2016. Global Status Report Renewables 2016 Global Status Report. Paris, http://www.ren21.net/status-of-renewables/global-status-report/.

REN21, 2018. Global Status Report Renewables 2018 Global Status Report. Paris. http://www.ren21.net/gsr-2018/chapters/chapter_03/chapter_03/.

Republic of South Africa (RSA), 2008. Regulations for the prohibition of the use, manufacturing, import and export of asbestos and asbestos containing materials. Government Regulation 341, Government Gazette 30904 of 28 March 2008.

Republic of South Africa, 2011. National Waste Information Baseline. Available at: http://sawic.environment.gov.za/documents/1880.pdf.

República de Colombia, Superintendencia de Servicios Públicos Domiciliarios. Informe Nacional de Aprovechamiento – 2016. Consulted on line in http://www.andi.com.co/Uploads/22.%20Informa%20de%20Aprovechamiento%20187302.pdf.

Resolución Cra 720 de, 2015. Ministerio de Vivienda, Ciudad y Territorio. República de Colombia. Comisión de Regulación de Agua Potable y Saneamiento Básico.

Riera, L., Wilson, B.W., 2015. Municipal solid waste as renewable resource. In: Presented at the 4th Global Economic Leaders' Summit, Changchun China, August 30–September 1. http://www.eprenewable.com/uploads/files/67_MSW-as-Resource-China_Final.pdf.

Rouse, M.W., 2005. Manufacturing practices for the development of crumb rubber materials from whole tires. (Chapter 1). In: De, S.K., Isayev, A.I., Khait, K. (Eds.), Rubber Recycling. CRC Taylor & Francis.

Rubber Manufacturers Association (RMA), 2016. 2015 U.S. Scrap Tire Management Summary. https://rma.org/sites/default/files/RMA_scraptire_summ_2015.pdf. (Accessed 30 April 2018).

Rubber Manufacturers Association (RMA), 2015. U.S. Scrap tire management summary. August 2016. https://rma.org/sites/default/files/RMA_scraptire_summ_2015.pdf. (Accessed 30 April 2018).

Ryckebosch, E., Drouilon, M., Vervaeren, H., 2008. Techniques for transformation of biogas to biomethane. Biomass Bioenergy 35, 1633–1645.

Sanches-Pereira, A., Tudeschini, L.G., Coelho, S.T., 2016. Evolution of the Brazilian residential carbon footprint based on direct energy consumption. Renew. Sust. Energ. Rev. 54, 184–201. http://linkinghub.elsevier.com/retrieve/pii/S1364032115009946.

Sanga, G., Jannuzzi, G.D.M., 2005. Impacts of Efficient Stoves and Cooking Fuel Substitution in Family Expenditures of Urban Household in Dar es Salam, Tanzania. Energy Discussion Paper no. 2.59.1/05, International Energy Initiative.

Sansuy, 2017. Indústria de Plásticos. Sem cheiro e aparência, Aterro Sanitário de Salvador torna-se referência. https://www.aecweb.com.br/emp/cont/m/sem-cheiro-e-aparencia-aterro-sanitario-de-salvador-tornasereferencia_13751_2694/. (Accessed 1 September 2017).

Scarlat, N., Motola, V., Dallemand, J.F., Monforti-Ferrario, F., Mofor, L., 2015. Evaluation of energy potential of municipal solid waste from Africa urban areas. Renew. Sust. Energ. Rev. 50, 1269–1286. https://doi.org/10.1016/j.rser.2015.05.067.

Scheinberg, A., Savain, R., 2015. Valuing Informal Integration: Inclusive Recycling in North Africa and the Middle East. Deutsche Gesellschaft für Internationale Zusammenarbeit (GIZ) GmbH, Echsborn, Germany. Available online at http://www.retech-germany.net/fileadmin/retech/03_themen/themen_informeller_sektor/Valuing_Informal_Integration.pdf.

Schnürer, A., Jarvis, A., 2010. Microbiological Handbook for Biogas Plants. Swedish Gas Centre Report 207, pp. 13–138. http://www.sciepub.com/reference/93407. (Accessed 23 July 2019).

Sebastian, R.M., Alappat, B., 2016. Thermal properties of indian municipal solid waste over the past, present and future years and its effect on thermal waste to energy facilities. Civ. Eng. Urban Plan. An Int. J. 3, 97–106. https://doi.org/10.5121/civej.2016.3208.

Sebastian, R.M., Kumar, D., Alappat, B., 2017. Comparative assessment of incinerability of municipal solid waste over different economies. In: Proceedings Sardinia 2017. 15th International Waste Management and Landfill Symposium Cagliari, Italy.

Sebastian, R.M., Kumar, D., Alappat, B., 2018. Comparative assessment of incinerability of municipal solid waste over different economies. Detritus 2, 89–95. https://doi.org/10.31025/2611-4135/2018.13658.

Sebastian, R.M., Kumar, D., Alappat, B.J., 2019a. Identifying appropriate aggregation technique for incinerability index. Environ. Prog. Sustain. Energy 38 (3), 1–10. https://doi.org/10.1002/ep.13068.

Sebastian, R.M., Kumar, D., Alappat, B.J., 2019b. An easy estimation of mixed municipal solid waste characteristics from component analysis. J. Environ. Eng. https://doi.org/10.1061/(ASCE)EE.1943-7870.0001588.

Sebastian, R.M., Kumar, D., Alappat, B.J., 2019c. Demonstration of estimation of incinerability of municipal solid waste using incinerability index. Environ. Dev. Sustain. https://doi.org/10.1007/s10668-019-00407-3.

Second Biennial Update Report of Argentina to the United Nations Framework on Climate Change Convention. 2017.

SEMARNAT, 2015a. Informe de la Situación del Medio Ambiente en México. Edición 2015. Capítulo.

SEMARNAT, S.d., 2015b. Intended Nationally Determined Contribution. SEMARNAT, Ciudad de México. Retrieved April 08, 2018.

SEMARNAT, S.d., 2016. Informe de la situación del medio ambiente en México. SEMARNAT, Ciudad de México. Retrieved March 15, 2018, from, http://apps1.semarnat.gob.mx/dgeia/informe15/.

SENER, S.d., 2017. Reporte de avance de energías limpias. 1er semestre 2017. SENER, Ciudad de México. Retrieved April 23, 2018.

Seo, Y., 2013. Current MSW Management and Waste-to-Energy Status in the Republic of Korea, pp. 1–65.

Shakya, P.R., Shrestha, P., Tamrakar, C.S., Bhattarai, P.K., 2008. Studies on potential emission of hazardous gases due to uncontrolled open-air burning of waste vehicle tyres and their possible impacts on the environment. Atmos. Environ. 42, 6555–6559.

Sharholy, M., Ahmad, K., Mahmood, G., Trivedi, R.C., 2008. Municipal solid waste management in Indian cities—a review. Waste Manag. 28, 459–467. https://doi.org/10.1016/j.wasman.2007.02.008.

Sharifah, A.S.A.K., Abidin, H.Z., Sulaiman, M.R., Khoo, K.H., Ali, H., 2008. Combustion characteristics of Malaysian municipal solid waste and predictions of air flow in a rotary kiln incinerator. J. Mater. Cycles Waste Manag. 10, 116–123. https://doi.org/10.1007/s10163-008-0207-3.

Sienkiewicz, M., Kucinska-Lipka, J., Janik, H., Balas, A., 2012. Progress in used tyres management in the European Union: a review. Waste Manag. 32, 1742–1751.

Simelane, T., Mohee, R., 2012. Future Directions of Municipal Solid Waste Management in Africa. AISA Policy Brief No 81. September 2012.

Singh, P., 2013. Impact of solid waste on human health: A case study of Varanasi city. Inter. J. Sci. Eng. Res. 4 (11), 1840–1842.

Singh, A., Spak, S.N., Stone, E.A., Downard, J., Bullard, R.L., Pooley, M., Kostle, P.A., Mainprize, M.W., Wichman, M.D., Peters, T.M., Beardsley, D., Stanier, C.O., 2015. Uncontrolled combustion of shredded tires in a landfill—part 2: population exposure, public health response, and an air quality index for urban fires. Atmos. Environ. 104, 273–283.

Smil, V., 2003. Energy at the Crossroads: Global Perspectives and Uncertainties. The MIT Press, Cambridge.

Smil, V., 2010. Energy Transitions: History, Requirements, Prospects. Praeger, Santa Barbaba, CA.

Smith, K., 2000. National burden of disease in India from indoor air pollution. Proc. Natl. Acad. Sci. USA 97, 13286–13293.

SNIS, 2016. Sistema Nacional de Informações Sobre Saneamento. http://app.cidades.gov.br/serieHistorica/.

Sokona, Y., Sarr, S., Wade, S., Togola, I., 2004. Energy Services for the Poor in West Africa: Sub-Regional "Energy Access" Study of West Africa. Global Network on Energy for Sustainable Development (GNESD).

Sokona, Y., Mulugetta, Y., Gujba, Y., 2013. Widening energy access in Africa: towards energy transition. Energy Policy 47, 3–10. https://doi.org/10.1016/j.enpol.2012.03.040.

Soliano Pereira, O.L., 1992. Rural Electrification and Multiple Criteria Analysis: A Case Study of the State of Bahia, in Brazil. Ph.D. Thesis, Imperial College of Science, Technology and Medicine, University of London, London.

Solomon, A.O., 2011. The role of households in solid waste management in East Africa capital cities. PhD thesis, Wageningen University.

Solvi, 2010. Mesa redonda "Mercado de Metano" – Experiência Brasileira do Grupo Solvi com Gás Metano. Available in https://pubs.naruc.org/pub.cfm?id=538DC7E5-2354-D714-516E-5D83415E58F8. (Accessed March 2018).

Srivastava, V., Ismail, S.A., Singh, P., Singh, R.P., 2015. Urban solid waste management in the developing world with emphasis on India: challenges and opportunities. Rev. Environ. Sci. Biotechnol. 14, 317. https://doi.org/10.1007/s11157-014-9352-4.

Stafford, F.N., Viquez, M.D., Labrincha, J., Hotza, D., 2015. Advances and challenges for the co-processing in Latin American cement industry. Procedia Mater. Sci. 9, 571–577.

State Grid, 2018. http://www.sgcc.com.cn/ywlm/index.shtml. (Accessed 22 July 2019).

State of Green, 2018. https://stateofgreen.com/en/profiles/babcock-and-wilcox-voelund/solutions/world-s-largest-waste-to-energy-power-plant. (Accessed 15 April 2018).

Sustainable Cities Programme (SACN), 2014. A Case for Municipal Solid Waste Management in Analysing Cities Financial Implication of Transitioning to a Green Economy. Available at http://www.sacities.net/wp-content/uploads/2014/08/waste_chapter_on_green_economy.pdf.

Tabata, T., Tsai, P., 2016. Heat supply from municipal solid waste incineration plants in Japan: current situation and future challenges. Waste Manag. Res. 34, 148–155. https://doi.org/10.1177/0734242X15617009.

Tahvonen, O., Salo, S., 2001. Economic growth and transitions between renewable and nonrenewable energy resources. Eur. Econ. Rev. 45 (8), 1379–1398.

Talyan, V., Dahiya, R.P., Sreekrishnan, T.R., 2008. State of municipal solid waste management in Delhi, the capital of India. Waste Manag. 28, 1276–1287.

Tangri, N., Shah, D., 2011. Third World Network, CDM Misadventures in Waste Management. June 9. https://www.no-burn.org/cdm-misadventures-in-waste-management/. (Accessed 22 July 2019).

Tanner, V.R., 1965. Die Entwicklung der Von-Roll-Müllverbrennungsanlagen (the development of the Von-Roll incinerators). Schweizerische Bauzeitung 83, 251–260.

Tchobanoglous, G., Theisen, H., Vigil, S.A., 1993. Integrated Solid Waste Management: Engineering, Principles and Management Issues. McGraw-Hill International Editions.

Technological Needs Assessment for Climate Change TNA, 2012. Final Report on Technologies for Mitigation. http://www.tech-action.org/Participating-Countries/Phase-1-Latin-America-and-the-Caribbean/Argentina.

Tello, P., Martínez, E., Daza, D., Soulier, M., Terraza, H., 2010. Informe de la Evaluación Regional del Manejo de Residuos Sólidos Urbanos en America Latina y el Caribe. OPS-AIDIS-BID.

Themelis, N., Diaz, M.E., Estevez, P., Gaviota, M., 2013. Guía para la Recuperación de Energía y Materiales de Residuos. Original study sponsored by Banco Interamericano de Desarrollo, Edición en español, 2016. Universidad del Desarollo, Chile; Columbia University, USA.

Third World Academy of Sciences (TWAS), 2008. Sustainable Energy for Developing Countries. TWAS, Trieste.

TNA/MAE/URC/GEF, 2016. Evaluación de Necesidades Tecnológicas para la Generación de Energía a partir de Residuos Sólidos Urbanos - Mitigación - Sector Energía - Ecuador Country Report. Ministerio del Ambiente MAE, Quito, Ecuador.

Tolmasquim, M.T., 2003. Fontes Renováveis de Energia no Brasil. Editora Interciência, Rio de Janeiro.

Traeger, R., et al., 2017. ALDC: The Least Developed Countries Report 2017: Transformational Energy Access. Geneva.

Training and Research Support Centre (TARSC), 2010. Waste Solid Waste Disposal in Three Local Authorities in Zimbabwe.

Tsiko, R.G., Togarepi, S., 2012. A situational analysis of waste management in Harare, Zimbabwe. Am. J. Sci. 8 (4), 692–706.

Tumuhairwe, J.B., Tenywa, J.S., Otabbong, E., Ledin, S., 2009. Comparison of four low-technology composting methods for market crop wastes. Waste Manag. 29 (8), 2274–2281. https://doi.org/10.1016/j.wasman.2009.03.015.

Tumwesige, V., Fulford, D., Davidson, G.C., 2014. Biogas appliances in Sub-Sahara Africa. Biomass Bioenergy 70, 40–50. https://doi.org/10.1016/j.biombioe.2014.02.017.

Tumwesige, V., Harroff, L., Apsley, A., Semple, S., Smith, J., 2015. Feasibility Study Assessing the Impact of Biogas Digesters on Indoor Air Pollution in Households in Uganda. In: Proceedings of the International Conference, University of California, Berkeley, p. 64.

Umeki, E.R., de Oliveira, C.F., Torres, R.B., dos Santos, R.G., 2016. Physico-chemistry properties of fuel blends composed of diesel and tire pyrolysis oil. Fuel 185, 236–242.

UN, 2018. United Nations Statistics Division. Retrieved July 27, 2018, from, http://data.un.org/DocumentData.aspx?q=HDI&id=377.

UN AGECC, 2010. Energy for a Sustainable Future. New York.

UNDP-HDI, 2016. Human Development Report 2016: Human Development for Everyone. Available from http://hdr.undp.org/sites/default/files/2016_human_development_report.pdf.

UNECA, 2009. Sixth Session of the Food Security and Sustainable Development. Africa Review Report on Waste Management – Main Report, Addis Ababa, Ethiopia, 27–30 October 2009. Available at:http://www.uneca.org/publications/africa-review-report-waste-management-main-report.

UNEP, 2013. Guidelines for National Waste Management Strategies: Moving from Challenges to Opportunities. United Nations Environment Program.

UNEP, 2015. Global Waste Management Outlook (GWMO). UNEP DTIE International Environmental Technology Centre, Osaka.

UNEP, 2016. Análisis de las (I)NDC de la región de América Latina y el Caribe. http://www.latinamerica.undp.org/content/rblac/es/home/library/environment_energy/analisis-de-las-i-ndc-de-la-region-de-america-latina-y-el-carib.html.

UNEP, 2018. Africa Waste Management Outlook. United Nations Environment Program.

UNEP (United Nations Environment Programme), 2004. State of Waste Management in South East Asia. http://www.aseansec.org/files.unep.pdf (Retrieved November 2011).

UNFCCC, 2015. Ecuador's Intended Nationally Determined Contribution (INDC).

UNFCCC, 2017. Biennial update report (BUR). BUR 2. Argentina.

UN-Energy, 2007. Energy for Sustainable Development: Policy Options for Africa. UN-ENERGY/Africa. UN-ENERGY/Africa Publication to CSD15, United Nations. Available at http://www.un-energy.org/sites/default/files/share/une/efsdpofa.pdf.

UN-Habitat, 2010. Solid Waste Management in the World's Cities. Water and Sanitation in the World's Cities. UN-Habitat, Nairobi. Available online at.

UN-Habitat, 2013. Solid Waste Management Assessment Within Urban Settings in Kenya and Uganda. UN Habitat, Nairobi, Kenya.

United Nations Conference on Trade and Development (UNCTAD), United Nations Environment Programme (UNEP), 2008. Organic Agriculture and Food Security in Africa. Available from: http://www.unctad.org/en/docs/ditcted200715_en.pdfS.

United Nations Development Programme (UNDP), World Health Organisation (WHO), 2009. The Energy Access Situation in Developing Countries: A Review Focusing on the Least Developed Countries and Sub-Saharan Africa. Available from: http://content.undp.org/go/cms-service/stream/asset/?asset_id=2205620S.

United Nations Industrial Development Organisation (UNIDO), 2009. Scaling up renewable energy in Africa. In: Proceedings of the 12th Ordinary Session of Heads of State and Governments of the African Union, Addis Ababa, Ethiopia. Available from: http://www.uncclearn.org/sites/www.uncclearn.org/files/unido11.pdfS.

Urban Research and Training Consultancy E.A Limited (URTC), 2012. Proposed Garbage Compositing Plant for Hoima Municipal Council. Environmental Impact Statement (EIS), Hoima, Uganda.

USAID (United State Agency International Development), 2006. Comparative Assessment: Community Based Solid Waste Management (CBSWM), Medan, Bandung, Subang and Surabaya. November 2006.

USEPA, 2009. Resource Assessment for Livestock and Agro-Industrial Waste—Argentina. Methane to Markets Program.

USEPA, 2017. Combined Heat and Power Partnership, Catalog of CHP Technologies. U.S. Environmental Protection Agency.

Van der Plas, R.J., Abdel-Hamid, M.A., 2005. Can the woodfuel supply in sub-Saharan Africa be sustainable? The case of N'Djaména, Chad. Energy Policy 33 (3), 297–306.

Van Dijk, M.P., Oduro-Kwarteng, S., 2007. Urban management and solid waste issues in Africa. In: A Contribution to the ISWA World Congress in September 2007 in Amsterdam.

Veolia, 2017. Veolia. Retrieved April 19, 2018, from Veolia convertirá en energía los residuos de la Ciudad de: https://www.veolia.com.mx/sites/g/files/dvc156/f/assets/documents/2017/04/Boletin_de_prensa_TermoMX_21.04.pdf.

Visvanathan, C., Glawe, U., 2006. Domestic solid waste management in South Asian countries: a comparative analysis. In: Presented Paper at 3R South Asia Expert Workshop, 30 August–1 September 2006, Kathmandu, Nepal.

Vögeli, Y., Lohri, C.R., Gallardo, A., Diener, S., Zurbrügg, C., 2014. Anaerobic Digestion of Biowaste in Developing Countries. EAWAG, Dübendorf.

Wade-Murphy, J., 2018. Abatement Costs in the Waste Sector of Ecuador. CAF.

Walekhwa, P.N., Lars, D., Mugisha, D., 2014. Economic viability of biogas energy production from family-sized digesters in Uganda. Biomass Bioenergy 70, 26–39. https://doi.org/10.1016/j.biombioe.2014.03.008.

Wang, Z., Li, K., Lambert, P., Yang, C., 2007. Identification, characterization and quantitation of pyrogenic polycylic aromatic hydrocarbons and other organic compounds in tire fire products. J. Chromatogr. A 1139, 14–26.

Waste Management Outlook for India. Consulted on line in BioEnergy Consult. 2016. https://www.bioenergyconsult.com/swm-outlook-india/.

Waste-to-Energy in China: Perspectives, 2016. Consulted on line in BioEnergy Consult. https://www.bioenergyconsult.com/waste-to-energy-china/.

WaterAid Uganda, 2011. From Rhetoric to Reality! Challenges of Urban Solid Waste Management in Kampala City. Issues for Policy Change.

WB, 2015. Beyond Connections Energy Access Redefined. Technical Report 008/15, ESMAP, Sustainable Energy for All.

Wb, T.W., 2018a. The World Bank. Retrieved March 31, 2018, from GDP per capita (current US$): https://data.worldbank.org/indicator/NY.GDP.PCAP.CD?end=2016&start=1960&view=chart.

WB, T.W., 2018b. The World Bank. Retrieved April 13, 2018, from GDP, current (US$): https://data.worldbank.org/indicator/NY.GDP.MKTP.CD?end=2016&start=1960&view=chart.

WEC, 2010. 2010 Survey of Energy Resources. World Energy Council, London, pp. 360–370. https://www.worldenergy.org/wp-content/uploads/2012/09/ser_2010_report_1.pdf. (Accessed 23 July 2019).

WEF, 2018. The Global Competitiveness Report: Chapter 2. World Economic Forum. Available at: http://www3.weforum.org/docs/GCR2018/02Chapters/Chapter%201.pdf.

Wellinger, A., Murphy, J.P., Baxter, D., 2013. The Biogas Handbook: Science, Production and Applications. Elsevier, Cambridge.

What a Waste: Solid Waste Management in Asia. World Bank. 1999.

What a Waste: Solid Waste Management in Asia. World Bank. 2012.

Wheles, E., Pierce, J., 2004. Siloxanes in landfill and digester gas update. In: 27th Annual SWANA LFG Symposium.

Williams, P.T., 2013. Pyrolysis of waste tyres: a review. Waste Manag. 33, 1714–1728.

Willumsen, H., 2001. Energy Recovery from Landfill Gas in Denmark and Worldwide. LG Consultant.

Wingqvist, Ő.G., Slunge, S., 2013. Governance Bottlenecks and Policy Options for Sustainable Materials Management—A Discussion Paper. United Nations Development Programme and the Swedish Environmental Protection Agency.

World Bank, 2008. Declining Rural Poverty Has Been a Key Factor in Aggregate Poverty Reduction. Focus A. Available from: http://siteresources.worldbank.org/INTWDR2008/Resources/2795087-1192112387976/WDR08_03_Focus_A.pdfS.

World Bank, 2009. Africa energy poverty. In: Proceedings of the G8 Energy Ministers Meeting 2009 Rome, May 24–25, 2009. Available from: http://www.g8energy2009.it/pdf/27.05/G8_Africa_Energy_Poverty_May9_Final_JS_clean.pdfS.

World Bank, 2011a. World Development Indicators. Available from: http://data.worldbank.org/data-catalog/world-development-indicatorsS.

World Bank, 2011b. Enterprise Surveys Database. Washington, DC. Available from: http://enterprisesurveys.org/DataS.

World Bank, 2016. Service Level Benchmarking for Urban Water Supply, Sanitation and Solid Waste Management in Zimbabwe: Peer Review Annual Report—2015 (English). World Bank Group, Washington, DC. Available on http://documents.worldbank.org/curated/en/661941477992625408/Service-level-benchmarking-for-urban-water-supply-sanitation-and-solid-waste-management-in-Zimbabwe-peer-review-annual-report-2015. (Accessed 20 February 2018).

World Bank, 2017. PIB (US$ a precios actuales) - Datos del Banco MundialWorld Bank. Available in https://datos.bancomundial.org/indicator/NY.GDP.MKTP.CD?locations=EC. (Accessed 2 April 2018).

World Bank, 2018. https://data.worldbank.org/country/japan. (Accessed 18 April 2018).

World Development Indicators, 2016. Highlights. http://databank.worldbank.org/data/download/site-content/wdi-2016-highlights-featuring-sdgs-booklet.pdf. 8 October 2017).

World Energy Council, 2016. World Energy Resources: Waste to Energy. World Energy Council, UK.

World Health Organisation (WHO), 2005. Indoor Air Pollution, Health and the Burden of Disease: Indoor Air Thematic Briefing 2. World Health Organisation, Geneva.

World Health Organisation (WHO), 2009. Global Health Risks. Mortality and Burden of Disease Attributable to Selected Major Risks. Available from, http://www.who.int/healthinfo/global_burden_disease/GlobalHealthRisks_report_full.pdfS.

Worrell, E., Price, L., Martin, N., Hendriks, C., Ozawa Meida, L., 2001. Carbon dioxide emissions from the global cement industry. Annu. Rev. Energy Environ. 26, 303–329.

WTERT—University of Columbia, http://www.seas.columbia.edu/earth/wtert/.

Xin-Gang, Z., Gui-Wu, J., Ang, L., Yun, L., 2016. Technology, cost, a performance of waste-to-energy incineration industry in China. Renew. Sust. Energ. Rev. 55, 115–130. https://doi.org/10.1016/j.rser.2015.10.137.

Xu, S., He, H., Luo, L., 2016. Status and prospects of municipal solid waste to energy technologies in China. In: Karthikeyan, P., Heimann, K., Muthu, S.S. (Eds.), Recycling of Solid Waste for Biofuels and Bio-chemicals. Springer. pp. 31–54. https://doi.org/10.1007/978-981-10-0150-5. ISBN 978-981-10-0148-2.

Yadav, P., Samadder, S.R., 2017. A global prospective of income distribution and its effect on life cycle assessment of municipal solid waste management: a review. Environ. Sci. Pollut. Res. 24, 9123–9141. https://doi.org/10.1007/s11356-017-8441-7.

Yamamoto, O., 2002. Solid waste treatment and disposal experiences in Japan. In: Proc. Int. Symp. Environ. Pollut. Control Waste Manag, pp. 417–424.

Yan, J.H., Chen, T., Li, X.D., et al., 2006. Evaluation of PCDD/Fs emission from fluidized bed incinerators co-firing MSW with coal in China. J. Hazard. Mater. A135, 47–51.

Yhdego, M., Kingu, A., 2017. The Anatomy of Solid Waste Mismanagement in East African Cities: Case of Dar es Salaam, Kampala and Nairobi. Environmental Resources Consultancy, Dar es Salaam, Tanzania.

Yokoyama, T., Suzuki, Y., Akiyama, H., 2001. Improvements and recent technology for fluidized bed waste incinerators. NKK Tech. Rev. 85, 38–43.

Yolin, C., 2015. Waste Management and Recycling in Japan Opportunities for European Companies (SMEs focus) 142.

Yoshida, H., Takahashi, K., Takeda, N., Sakai, S.i., 2009. Japan's waste management policies for dioxins and polychlorinated biphenyls. J. Mater. Cycles Waste Manag. 11, 229–243. https://doi.org/10.1007/s10163-008-0235-z.

ZERA, 2016. Zimbabwe Energy Regulatory Authority, 2016 Annual Report.

Zhang, D.Q., Tan, S.K., Gersberg, R.M., 2010. Municipal solid waste management in China: status, problems and challenges. J. Environ. Manag. 91 (8), 1623–1633. https://www.sciencedirect.com/science/article/pii/S0301479710000848. (Accessed 22 July 2019).

Zhang, D., Huang, G., Xu, Y., Gong, Q., 2015. Waste-to-energy in China: key challenges and opportunities. Energies 8, 14182–14196. https://doi.org/10.3390/en81212422.

Zheng, L., Song, J., Li, C., Gao, Y., Geng, P., Qu, B., Lin, L., 2014. Preferential policies promote municipal solid waste (MSW) to energy in China: current status and prospects. Renew. Sust. Energ. Rev. 36, 135–148. https://doi.org/10.1016/j.rser.2014.04.049.

ZimStat, 2016. Zimbabwe Water and Statistic Report-2016.

Zurbrugg, C., 2002. Urban solid waste management in low-income countries of Asia: how to cope with the garbage crisis. In: Presented for: Scientific Committee on Problems of the Environment (SCOPE), Urban Solid Waste Management Review Session, Durban, South Africa, November 2002.

Zurbrugg, C., Drescher, S., Patel, A., Sharatchandra, H.C., 2004. Decentralised composting of urban waste—an overview of community and private initiatives in Indian cities. Waste Manag. 24, 655–662.

Index

Note: Page numbers followed by *f* indicate figures and *t* indicate tables.

Printed in the United States
By Bookmasters